The Joy of Flying

3rd Edition

Robert Mark

TAB Books

Imprint of McGraw-Hill

New York San Francisco Washington, D.C. Auckland Bogotá
Caracas Lisbon London Madrid Mexico City Milan
Montreal New Delhi San Juan Singapore
Sydney Tokyo Toronto

pbk   4 5 6 7 8 9 10 11 DOC/DOC 0 9 8 7 6 5 4 3 2 1 0

Library of Congress Cataloging-in-Publication Data

Mark, Robert (Robert Paul)
    The joy of flying / by Robert Mark.—3rd ed.
        p.   cm.
    Includes index.
    ISBN 0-07-040487-9
    1. Private flying. 2. Airplanes, Private. I. Title.
TL721.4.M37 1993
629.132'5217—dc20   93-33783
                                                              CIP

Acquisitions Editor: Jeff Worsinger
Editorial team: Joanne Slike, Executive Editor
                        Susan Wahlman Kagey, Associate Managing Editor
                        Charles Spence, Editor
                        Joann Woy, Indexer
Production team: Katherine G. Brown, Director
                        Sue Kuhn, Typesetting
                        Nancy Mickley, Proofreading
                        Wendy Small, Layout
Design team: Jaclyn J. Boone, Designer
                        Brian Allison, Associate Designer
Cover photograph: Bender and Bender Photography, Waldo, Oh.   AV1
Cover design: Denny Bond, East Petersburg, Pa.   4412

For Nancy

Contents

8   Time to move up?   131

9   Building your own airplane   147

10   Some final thoughts   157

Index   163

"WE CAN LIFT OURSELVES out of ignorance, we can find ourselves as creatures of excellence and intelligence and skill. We can be free. We can learn to fly!"

RICHARD BACH
Jonathan Livingston Seagull

Introduction

ALL YEAR LONG, I TRAVEL AROUND THE COUNTRY ON BUSINESS, much of it in a general aviation aircraft. No matter where I end up, the subject of how I arrived at my destination always seems to become a topic of conversation. Even as we approach the twenty-first century, at a time when millions of people around the globe rely on air transportation as a common form of transportation, people are amazed at the thought that real, live, everyday people can pilot airplanes around the country just as professional pilots do.

But while clients are certainly impressed when they pick me up at the airport after I shut down the engine of the Cessna 172RG I fly, they are even more impressed when they learn that they, too, could be flying such an airplane in a great deal less time than they ever imagined. At first, learning to fly an airplane might bring to mind that old joke, "How do you eat an elephant? Why, one piece at a time, of course!" The first time one of my clients looks inside the cockpit of my airplane and fixates on the instrument panel, the wide-eyed question is usually, "But how do you remember what all those dials and knobs are for?" The answer, again, is really simple. Your flight training begins with just one very small step—taking an introductory ride of a half hour or so. During this flight, your instructor flies as he or she explains in very general terms what the various flight instruments actually tell the pilot. "Here is the altimeter that shows how high we are, and right now it's indicating 2,500 feet. Here is the airspeed indicator that shows we're flying through the air at 95 knots, or about 103 mph," and so on. During this first demonstration flight, the instructor usually tells students to put their hands on the control wheel and feet on the rudders and follow through with the instructor during a few gentle turns, climbs, and descents. This way new pilots gain a feel for what is happening as those maneuvers are executed. By the time students try a few turns on their own, big smiles appear on their faces that last long after the aircraft has returned to the airport and been tied down.

The thought in your mind right now is, "Sure, but I could never learn to fly because I'm not that good at math" or "I really don't have a very good sense of direction." Nonsense. Just about anyone can learn to fly. Recently, an 11-year-old girl made a cross-country flight from the East Coast to the West Coast in three days with her instructor. She did all the flying herself. Believe me, it will only take one flight to convince you that you too can learn to fly. I already know you want to or you wouldn't be reading this book.

So stop filling your head with all sorts of reasons why you *can't* learn to fly. Inside these pages you'll find yourself bombarded with reasons why you *should* learn to fly. After you read this manual, don't put off making your dreams come true. Pick up the phone, dial the number of the nearest flight school, and tell them you'd like to take an introductory flight.

Once you find out how different life can be a few thousand feet above the earth, once you've been introduced to the possibility of making weekend flights to fairly remote, often romantic hideaways, you'll never want to stop. If that idea becomes reality for you, then this book and I will have accomplished our mission.

Good luck and happy flying.

1

The joy of flying

HARDLY A PERSON on the face of the earth has not watched a bird fly around his or her backyard or land on a fence post. Some people see them as just birds, hopping across the grass in search of breakfast while other people notice nothing at all as these creatures soar above their heads. There are people, though, who see magic in these moments. These people feel something when they see a flight of Canadian geese begin their final approach to a pond, or observe a flock of ducks begin their takeoff run across a lake and lift into the air, or just watch a hawk circle endlessly above the side of a road on a warm summer's day.

These curious souls aren't ornithologists spending hours categorizing birds. On the contrary, they're just ordinary folks, carpenters, housewives, and police officers, who don't just watch events, but feel them. But why do these people feel such an incredible affinity with the birds? Why bother to watch this creature's movements unless they plan to sketch them? The answer is simple. These people all share a common desire: They all wish they could fly.

Soon, you'll be able to experience the beauty of flight and share that vision with your family and friends. Millions of people turn their heads skyward at the first drone of a propeller-driven aircraft or the rush of air as a jet moves overhead and then disappears from sight. They sigh and go back to work, wishing they could join the fraternity of fliers. Imagine the joy they will feel when they learn they can.

There's an essence, something almost magnetic about airplanes. Even small children look up as an airplane passes magically overhead, unable to decipher what they see, yet fascinated nonetheless. Why are earthbound creatures so enamored with flight?

In a word, flight is freedom (Fig. 1-1). It gives the pilot and passengers the opportunity to view the earth from a vantage point not available to everyone. No longer is the earth just a farmer's field with a few scattered houses or a huge megapolis of building tops that soar above, sometimes out of sight. From an airplane, the perspective of earth we've known becomes a collage of tiny spectacular sights visible only from that special observation site 2000 feet above the ground. Unlike a normal observatory, this affords a constantly changing point of view. No

Fig. 1-1. Flying is freedom to go anywhere you choose.

two flights are ever the same nor any two views of the Manhattan or Chicago skyline ever alike.

Flying brings a sense of accomplishment for a pilot. Six months before, this man or woman was looking inside the cockpit of a small airplane, wondering in amazement at how a pilot could keep the meaning of all those dials, gauges, and switches straight. Now that same man or woman, a licensed aviator, has the ability to see why so much time and work are spent training to become a pilot. Certainly there were times when the going was tough, when they couldn't get the hang of landings. They scared themselves silly the first time they thought they were really lost on a cross-country flight.

Now that's all past and they've proven to a Federal Aviation Administration (FAA) flight examiner they have learned how to control this flying machine. They've proven they are capable of studying weather and route conditions for the flight, battling the wind if necessary to reach that point in the sky where everything is smooth. The pilot watches the hours drift by as the miles pass beneath the airplane, en route to some exotic vacation or business meeting. Maybe they're aloft simply for the opportunity to fly again. They pat themselves on the back for an outstanding job of learning the skills of an aviator or just to get away from it all at the end of a tough day.

They truly are free (Fig. 1-2). Aloft there are no road signs, no police officer hiding around that next clump of trees with radar equipment. Private pilots have the ability to take an airplane from any airport in the country and land just about anywhere they'd like. They are free. As free to choose their destination or change it at will as that blue jay fluttering around their yard: free to come and go. While they're aloft most pilots think: *If only the rest of the people on earth could see how beautiful it can be to leave the ground, if they could only see their city, their state, their own home from an airplane. If they knew what splendid view an airplane affords, they'd be up here too.*

Fig. 1-2. Aircraft come in all shapes and sizes.

Thousands of people have learned to fly in the past 50 years, but the current pilot population represents only a fraction of those people that are capable of learning. Why would anyone want to take the time and spend the money to learn how to do something that still seems so dangerous to some people? Part of the answer lies in our fear of what we don't understand. Consequently, some people allow their fear of flying, or fear of failing, or fear of something going wrong up there, to stifle a desire to learn what it's like to soar like a bird.

Some people are lucky because they're able to set their fears or doubts aside and look at learning to fly for what it is: a chance to experience something exciting. However, still too many people view learning to fly only from a dollars-and-cents standpoint. That truly is a mistake—not that the money for this investment isn't something to consider. Money, or the lack of it, shouldn't stop you from flying an airplane because, on the average, the goal of flying an airplane can be accomplished for less than 10 dollars a day for the 6 to 12 months it takes most people to learn to fly. (After that, the cost can be much less, as described in Chapter 2.) For now, forget the idea that flying costs too much. Let your mind run free with thoughts of soaring above the earth while you are in command of an airplane.

Still not sure? How about a no-obligation opportunity to check the entire program out and not spend a lot of money in the process? When you visit your local airport, you can take a demonstration ride for about 50 bucks. For 40 minutes, you'll be right up front, in the cockpit of an airplane, with a qualified instructor next to you showing you how everything works. Demo ride day isn't the time to try and learn how all the dials and gizmos work. The demo ride is the day you spend your time in the cockpit, looking out the windows, watching the earth from above, watching that red-tailed hawk circling beneath you off the left wing. Today is your chance to see and feel and experience what it's really like to be up there closer to the clouds. This is your chance to ask yourself why you've been putting this off for so long. It's the chance to tell yourself that you're not getting any younger. By the time you return to earth, you'll realize just how tough it will be to wait for that first flying lesson (Fig. 1-3).

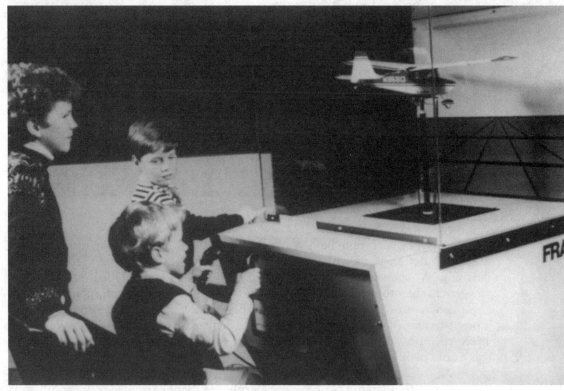

Fig. 1-3. Future pilots learn about flying in a Frasca basic flight simulator.

Perhaps you're a romantic. You believe in all the lore about flying. You know if you had the chance you'd don a leather cap and goggles. Maybe, too, you've always secretly admired Charles Lindbergh and his solo flight across the Atlantic, or Amelia Earhart's spectacular flights across both oceans. Come on . . . do you really need a cut-and-dry, practical reason for spending the time and money to learn how to fly?

In 1977, 243 million people flew from the major United States air carrier airports. In 1987 the figure was 468 million, and by the turn of the century, the figure will climb to 745 million. What does all this mean to the business traveler? Quite simply, it means you're going to encounter some incredible delays when flying through La Guardia, O'Hare, Denver, Los Angeles and the 28 other major hub airports around the country. In fact, the FAA believes by the year 2000 those delays will number 20,000 hours per year. Why not learn to fly and avoid the congestion? Why not learn to fly and stop wasting time?

If you're a business traveler who feels you spend your life in your car, I'm surprised you never learned to fly. It takes only a few moments to sit down with a pen and paper to show how much time it saves flying to sales calls and renting a car, instead of driving everywhere. Flying expenses are as deductible to you in your business as automobile costs (Fig. 1-4). For instance, depart Chicago at 8 A.M., arrive Madison, Wisconsin, 9 A.M., have a two-hour meeting, and fly to Springfield, Illinois, arriving at 1:15 P.M. You have a business lunch and tour the customer's factory and leave by 4 P.M., landing at Chicago by 5:15 P.M. This day you've had two meetings and covered 525 miles of travel and made it home in time for dinner.

Try that in your car and you'll be gone for days. But it's not much better on the airlines if you fly from Chicago Midway Airport to Madison, fly back to Midway to catch the plane to Springfield, then return. It's almost impossible to accomplish the same amount of work in one day flying the airlines as it would be flying an airplane of your own.

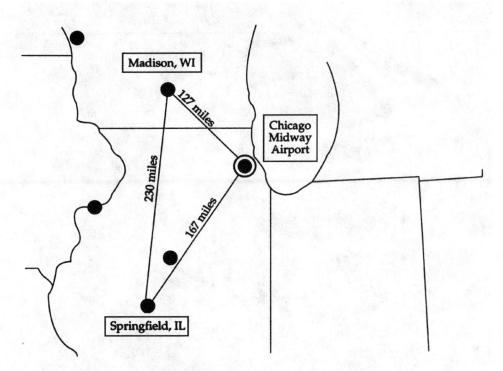

Fig. 1-4. An airline trip might involve legs from Midway to Madison, back to Midway, then to Springfield, and a return to Midway. General aviation is direct.

Certainly there are people who see only that practical need to fly on business. Others learn to fly because they've always felt that magical attraction of the skies. Some people fly as a means to an end, such as a career in aviation. As a professional pilot, you not only have the chance to fly often, but you get paid for it.

Sure, you'll run into people who will say that it's a mistake to turn something you like into something to make money. You'll lose something of your love in the process. I think they're wrong. I've been a pilot for 26 years, being paid for 17 of those years. Today, the feeling I get as I climb to altitude on a chilly morning and watch the shoreline at Chicago disappear behind me is still awe-inspiring.

I see it like no one else will. Only I am in that particular spot as the sun rises. Only I can catch the shadows of the morning on the ice below. I can see the city behind and the western shore of the state of Michigan ahead. How could anyone *not* want to share this experience? This is what flying is all about. It still gets to me after 26 years, just the way it did when I saw similar sights as a student pilot.

People who love what they do, whether it's work or play, are better at it than those who just work a job for the money or play tennis to make a good impression on the boss or business clients. I can't think of a better, more satisfying way to earn a living than flying. If your interest is just for something new to try, I can't imagine anything other than flying for the excitement, the chance for adventure, and the chance to explore the world (Fig. 1-5).

Fig. 1-5. Dale Crites flying his 1912 Curtis Pusher "Silver Streak" during the annual EAA Oshkosh convention.

WHAT DOES A PILOT REALLY DO?

What a pilot does seems almost obvious to most people: fly the airplane. For a moment, let's examine what pilots really do when they are sitting in the cockpit, and how they think as they fly.

Assume that the pilot has a private pilot certificate, single engine land airplane rating. This allows the pilot to fly a very typical aircraft, like the Cessna 172, containing four seats and room for baggage. It's powered by a 160-hp engine and cruises at about 115 to 120 mph. The airplane can carry enough fuel for a flight of about 3.5 hours, depending upon how heavy the rest of the load is. Very few airplanes can carry full fuel, full baggage, and fill all the cabin seats with people. It is usually a compromise to be certain that the design weight limits of the aircraft are not exceeded.

Late-model airplanes are usually made of metal and designed to withstand considerable physical stress. Most airplanes have been derived from a long family history that included improvements each year, as the Ford Thunderbird today is a much better automobile than it was 30 years ago. Major aircraft components are described in Fig. 1-6. While all the parts are important, the most visible are the propeller, the landing gear, the wings, the fuselage, the horizontal and vertical stabilizer, the elevators, and the rudder.

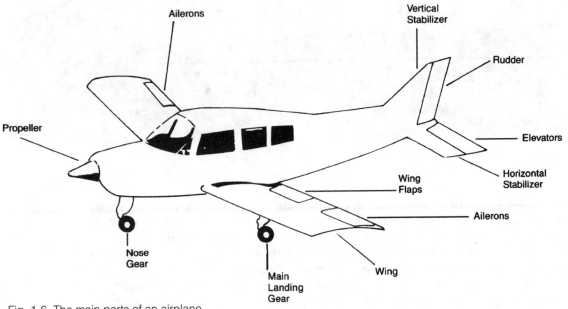

Fig. 1-6. The main parts of an airplane.

Inside the cockpit are gauges, switches, instruments, and electronics (Fig. 1-7). The basic flight instruments include:

- an altimeter to tell the pilot how high he is flying.
- an airspeed indicator to tell the speed of the plane through the air.

- an attitude indicator to represent the airplane's relationship to the horizon.
- a rate-of-turn indicator shows wing position and coordination.
- a heading indicator that, when properly adjusted, will do a better job than a standard magnetic compass.
- a magnetic compass.
- a vertical speed indicator, which shows rate of climb or descent.

Fig. 1-7. The cockpit of a Commander 114B.

Commander Aircraft Company

Below these instruments are the electrical switches. First is the master switch, which turns on all electrical power to the aircraft. Like an automobile, when the engine starts (using an onboard battery), an alternator attached to the engine supplies electrical power. Because an aircraft engine uses a magneto (a self-generating power unit) to supply the spark for the spark plugs, the entire electrical system could be turned off or fail in flight and the engine would continue to run. To play it safe, all aircraft engines have two sets of spark plugs and two magnetos, just in case one set fails. The aircraft's ignition and starter are controlled from a key switch.

Next to the ignition switch are the various controls for the navigation and landing lights as well as the interior lights. Lower on the panel are the circuit breakers that protect all the electric circuits in the plane.

In the middle of the instrument panel are the aircraft radios used for navigation and communication. An aircraft might have one to six radios. A small aircraft flown away from a major metropolitan area might contain no radio; this is the ultimate in flying freedom, but it does restrict flying to sparsely populated areas.

The main aircraft radio is called the nav/com, a combination of navigation and communications capabilities housed in a single unit. Because aircraft radios operate in the frequency band just above standard FM radio, 108 to 136 MHz, they are limited to line of sight reception. This means that the higher an aircraft is above the ground, the greater the range. For example, at 7000 feet the normal range could be about 50 miles.

Below the nav/com unit is an air traffic control (ATC) transponder. This unit electronically communicates with ground based ATC radar units to give controllers a better view of where you are. It also gives them a better chance to tell you about other air traffic that might be in your area. For normal flying, a transponder should be standard equipment in an aircraft and is required when flying in many busy areas of the country. The nav/com and a transponder are considered basic equipment on most machines these days. Many aircraft will carry two nav/com units. Another unit still found today in many aircraft is called an Automatic Direction Finder (ADF). This radio tunes in stations in the frequency band beginning at about 200 kHz through 1600 kHz, which is part of the AM broadcast band.

Beneath the radio stack, at the bottom of the panel are the engine controls. A long rod with a black knob on it is the throttle, which determines how fast the engine is going to run; all the way back is idle speed, all the way forward gives the engine full takeoff power that is transferred through to the propeller. A red knob on another rod is the engine fuel mixture control. Air becomes thinner as an aircraft climbs higher and this control allows the pilot to reduce the amount of fuel being mixed with the engine intake air. This mixture produces the best power available to the engine.

Engine gauges are on the right side of the panel. These include two fuel gauges, one for the right tank (located inside the right wing) and another for the left tank (on the other side of the aircraft). Additionally, oil pressure and oil temperature gauges give an indication of engine performance. The tachometer operates like in a car, telling the pilot how many revolutions per minute (rpm) the engine is turning. The propeller is attached directly to the crankshaft on this airplane, so the tach also reads propeller speed and allows the pilot to set the proper engine speed. The last two gauges are the electrical ammeter, which reveals charging system status, and the vacuum gauge, which tells whether or not certain flight instruments are operating properly.

The large control wheel is attached underneath the panel on the left side via a large piece of steel tubing just beneath the flight instruments. A second wheel is attached beneath the engine gauges on the right side of the instrument panel. On the ground, the control wheel has very little function, but in the air it is used in coordination with the pedals on the floor to move the ailerons, rudder, and elevator (Fig. 1-8). The rudder pedals have two functions on the ground: steering and braking. The top portion of the pedals are hinged to provide access to the brakes, one on either of the two main wheels.

Something you see in a great deal of airplanes is redundancy: two control wheels, two sets of rudder pedals, two magnetos. The main reason is safety: if one

Fig. 1-8. The cockpit of a Cessna 172 showing the rudder pedals, brakes, and fuel tank selector.

fails, there is a backup. Safety is always the major consideration for a pilot and as a student you will hear safety considerations expressed over and over. You as a private pilot will be the pilot-in-command of an aircraft and will be responsible for its operation. Operating that aircraft safely should be the most important thing on your mind when you fly, always.

The night before our typical cross-country flight of 260 miles, the pilot checks with the government weather office, called the Flight Service Station (FSS), to see just what kind of weather conditions might be expected on the flight. Another government service, called DUAT, as well as a few private services, allow a pilot to access weather information from home or office with a personal computer. If the weather is forecast to be bad late in the day, the pilot might decide to begin the journey early, or vice versa if the weather is bad at the beginning of the day. A flight plan is filed at the FSS indicating where the airplane is going, the route it will fly, and how long it will take to get there. If you do not arrive, a search will commence along the route.

The pilot plans the flight by obtaining the appropriate navigational charts, in this case, for the route from Chicago to Rhinelander, Wisconsin. Through the skills learned as a student, he or she can plot a course, allowing for winds, to make sure the aircraft lands in Rhinelander and not Minneapolis. The pilot also plans the fuel load and makes sure there is enough to get the aircraft to Rhinelander with plenty of reserve, just in case the weather deteriorates and landing at Rhinelander is prohibited.

When the passengers arrive at the airport, the pilot makes a thorough preflight check of the airplane to be certain everything is in perfect operating order before the engine is started. Are the tires in good shape? Is there any frost on the wings or any nicks in the propeller? He checks the fuel supply to be certain no water has collected in the tank during the night and that the engine oil is at the proper level.

After engine start, the pilot runs the engine up to nearly full power to make certain the oil pressure and rpm are within prescribed limits. If not, the pilot turns the aircraft around and heads for the shop for a mechanic to correct the problem.

Once the pilot is satisfied that all is correct, and he has received clearance from the control tower, the pilot lines the aircraft up on the runway, the throttle is pushed all the way in and the propeller turns faster, pulling the aircraft along the runway. When there is enough speed to keep enough air moving across the wings, the pilot pulls back on the control wheel, which moves the elevator at the tail of the aircraft, and the airplane leaves the ground.

After takeoff, the pilot checks his route of flight using some of the radio navigation aids in the plane and also compares what he sees on the ground to the items displayed on a chart (Fig. 1-9). Besides his navigation duties, the pilot is constantly looking out the windows to make sure the weather is holding up according to what the forecaster stated. If it isn't, perhaps a deviation around some bad weather will be necessary; that's the pilot's decision.

En route, the pilot is constantly checking the airplane's progress across the ground. He or she makes certain the fuel reserves are sufficient if the wind turns out to be stronger than planned. The pilot calculates the ground speed and produces an estimate for arrival at the destination. En route, the pilot might be in contact with ATC, which will point out other traffic in the area, if workload permits.

This all sounds like a great deal of work! While it does require attention, it doesn't require so much brain work that the pilot and passengers are unable to enjoy the beauty of the landscape beneath, be it a large city, the Mississippi River, or Mount Rushmore.

Using radio, the pilot again checks weather at the destination to make certain the flight can land. The pilot locates the airport and communicates with ATC or an advisory service for winds and the proper runway to use, and lands.

The pilot has saved three hours by flying and given the family an unparalleled sight-seeing trip. Time in the airplane, one hour, 57 minutes. By car, the same trip would have consumed nearly five and a half hours. The pilot might fly back later in the day, something that no one would consider doing in a car.

Those are some of the mandatory things a pilot does. What about some fun diversions? Maybe the pilot decided to stop at a resort overnight and go on to the destination the next morning; it's a quick trip in an airplane. Consider the businessman who's en route to Raleigh-Durham, North Carolina, from Miami and decides that because the weather is poor in Raleigh, he'll stop in Atlanta to see his sister and continue to North Carolina the next morning.

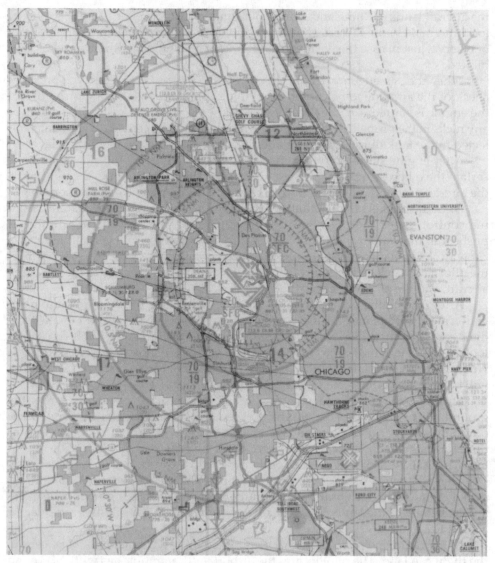

Fig. 1-9. Typical visual flight rules (VFR) navigation chart.

Fun trips include classics like Orlando, Florida, landing at Kissimmee and renting a car for the 10-minute drive to Disney World. Or perhaps spending the day at Disney World and hopping in the plane to head over to Tampa and Busch Gardens, a mere half-hour flight.

How about a flying tour of New England or the California coastline? The trouble with flying is that when your mind runs free, when all those little gremlins that have been saying that you're either too old or too poor to learn to fly are beaten back, you'll be bombarded with flying ideas. Eventually, you'll run out of time to fly everywhere.

Who knows what the motivation will be to stop at the airport and check into flying lessons? Whatever it is, don't let doubt stop you for another day. If you can borrow the money to learn to fly, the money you save by not smoking two packs of cigarettes a day would help repayment of that loan. Still think it's too late to try?

My friend Marge started telling me she wanted to learn to fly years before she actually got the chance. She'd flown with her husband in their airplane, but never got around to learning until she received a gift of flying lessons. That indoctrination started her on the way to what eventually became a commercial pilot license with instrument and flight instructor ratings. Flying began almost as a whim for Marge and she turned down an opportunity to attend medical school and continue flying as an instructor.

Certainly, one reason for flying is emotional because flying is just plain fun. In fact, some pilots might have no logical reason for learning to fly. Some have to overcome the objections of a spouse to accomplish the goal. To a spouse, learning to fly could be considered a total waste of money on something that is much too dangerous or much too complicated. This isn't uncommon.

It often takes time for a new pilot to sort it all out and decide whether or not he or she has the stamina to overcome objections of other people. Luckily, a great many people learn to fly with the total support of their families. They, too, see flying as much more than just something to do on Sunday afternoons.

OTHER THINGS TO FLY

Other flying machines, besides airplanes, can stir the interest of a new pilot. If you've never seen the launch of a hot air balloon, it can open your eyes to an almost mystical feeling of flight. At first glance, the entire hot air balloon—which can fit into a basket about four feet square—looks about as airworthy as a gas range. The ground crew that accompanies the balloon carefully unpacks it and unrolls the entire balloon on the ground. After lighting the propane burner that provides the hot air for the balloon, the crumpled lengths of cloth slowly begin to rise and take shape. Within a half hour, the balloon is standing alone, hovering over the small gondola, looking much like a giant multi-colored electric light bulb.

When preparations are complete, the pilot and passengers climb aboard. The burner cord is pulled to release more and more warm air into the balloon. Soon it begins to rise, taking the pilot and a passenger on a morning's flight above the countryside at whatever speed and direction the wind is producing at the moment.

Even more exciting than a single launch is a mass ascension: 6, 10, 15, 50 balloons, all sitting upright in a field ready to fly (Fig. 1-10). Large bags 30 feet high in blues, reds, greens, yellows, and even patterns, checkerboards, polka dots, and more, all waiting for their passengers.

Within minutes, the air is filled with balloons (Fig. 1-11). As a passenger, you see the land pass beneath you slowly. Silence is disturbed only occasionally by the roar of the gas burner. At 100 feet above the ground, you pass over children on bicycles. You can wave and be recognized, too. You can even yell to those below and be heard, yet you're free of the earth, you're flying.

Fig. 1-10. The 1988 Albuquerque International Balloon Fiesta takes shape.

Not to be outdone for pure smoothness and pure attraction to the skies is soaring (Fig. 1-12). It is also known as gliding. You fly an aircraft that has no engine! On the ground, a glider looks like an overgrown, gangly airplane. It has a short fuselage, with long wings—longer than most powered aircraft. A sailplane needs long wings to catch and fully utilize thermals, the invisible "up" elevators that are lifeblood to a sailplane. The relative calm of soaring adds a new sense of serenity to flying that is often missed in a powered airplane.

Another aspect of flying that doesn't require a separate rating and can open an incredibly exciting world to a pilot is aerobatics. Many pilots love aerobatics after the first time upside down. Aerobatics best suits those who love the thrill of a roller coaster, or even a fast-moving elevator. Various levels of participation exist, from the beginner who just wants to learn how to loop and roll an airplane, to the expert who learns the intricacies of Immelmans and Cuban 8s in competition. Every pilot that experiences aerobatics to any degree will come away from the experience with knowledge and a new respect and sense of confidence in both the airplane and the pilot's own flying abilities.

The 10 greatest lies to tell a spouse about flying

1. It's tax deductible.
2. It's cheap.

3. This will bring me closer to the kids.
4. I don't need any fancy equipment to learn.
5. I don't need any fancy flying clothes.
6. I'll always come straight home after the pilot association meetings.
7. I don't need a subscription to any of the aviation magazines.
8. I never get airsick.
9. I don't care if Joe soloed before me.
10. It's cheap.

Fig. 1-11. A balloon launch up close.

Paul Minter

NIGHT FLYING

The first time I flew at night was on one of my student pilot cross-country training flights around central Illinois. I was 17 at the time and had never experienced anything quite as awe inspiring.

We took off from Champaign, Illinois, en route to Springfield, Illinois, as the sun was setting on a calm, cloudless evening. Lights were just beginning to dot a

Fig. 1-12. A Schweizer 1-35 sailplane could introduce you to the realm of powerless flight.

landscape that I had come to know so well as a student; the ground below was only daytime landmarks spotted with light. As the darkness grew more intense, a curious thing happened. Very slowly, large familiar places on the ground disappeared from view, replaced in odd places by large groups of lights and big dark areas. From several thousand feet, I could see flashing red neon lights and headlights from automobiles snaking their way along the now invisible roads.

Each city we passed was covered in a soft white aura like a secret canopy designed to keep its residents safe. Below, the ground was alive with cars, busses, and billboards, as well as floodlights illuminating a baseball field. The flight visibility this night was superb. As we crossed over one town, we were able to see the next two or three towns highlighting the darkness as much as 75 miles ahead.

Flying toward Chicago another night, the lights on the ground were so brilliant, we saw the glow from more than 100 miles away. This aerial observatory also gave me the clearest view of the heavens I've had to this day.

An unrestricted private pilot license requires at least three hours of night flying experience. But three hours aloft, at night, is barely enough to grasp the beauty of this form of flight. Sometimes students will become so enamored with the experience that they only want to fly at night.

FLYING RELATIONSHIPS

Whether its skiing, playing golf, or flying, there is always a clubhouse or meeting place where friends get together for lunch or dinner and the chance to tell some "lies," as my friend Gordon Baxter calls them. These are the fish stories, the tales of the toughest flying day you've ever had, your report on the best airplane ever built, or your first flight in a different aircraft (Fig. 1-13). This is the chance to chat with other people who share your love of aviation, understand your stage of flying, or have experienced buying a first airplane (Fig. 1-14).

I really think it's important, as a potential pilot, to take advantage of these relationships at the airport, because a friend can always make tough times easier. Many of these people eat, sleep, and breathe airplanes. They know the anxiety of a

Fig. 1-13. A first flight in a helicopter can be a memorable experience.

Fig. 1-14. A Saturday meeting of the Palwaukee Airport lunch bunch.

difficult lesson (like your first landings) and many of them know what it's like to be short of cash. And it's so simple to build these relationships.

Ask your instructor to join you for a cup of coffee at the restaurant or offer to buy lunch. Tell your instructor that you'd like to meet other pilots and students. He or she can probably recommend organizations that meet locally, such as the local pilots association. Consider the Experimental Aircraft Association (EAA), which is devoted to many different facets of flying, such as building an airplane or learning about aerobatics. Consider EAA's Warbirds Division for those pilots who always wanted to get up close to those big planes that fought in World War II. How about volunteering for community work through the Civil Air Patrol, the civilian auxiliary of the U.S. Air Force. (See Chapter 8 for more on associations.) For the ladies there is the Ninety-Nines Association of women pilots.

People come to flying from many different places and for all sorts of reasons. Wendy remembers her early flying experiences with her father:

Flying seemed interesting to me, but not that much. My father was always trying to get me to take lessons but I just never did. Eventually, my mom and dad bought an airplane and when I turned 17 I started my training. I remember the first few lessons. They were OK, but I still wasn't plane crazy.

My love of flying came the day I made my first takeoff with my instructor next to me. As I pushed the throttle forward, it was hard to realize that the vibrations and power I felt surging through the aircraft were a result of what I was doing. We rolled down the runway and I pulled back on the control wheel.

The vibration stopped and I looked out the window and watched the shadow of the airplane getting smaller as we climbed away from the runway. I was flying. I was in control of where I was going . . . that's when it really hit me. I never stopped flying. I just couldn't ever.

Bob remembers being a passenger when his love for flying took hold:

I was in a small Cessna, I don't remember what kind, but my uncle was flying. We took off on a gloomy summer day with rain falling lightly against the windshield. To my amazement, we flew right into the clouds; my uncle had an instrument rating. Outside I saw only gray. The clouds were so thick that sometimes I couldn't even see the wheels of the airplane just a few feet away.

All of a sudden it started getting lighter above, not clear, not even breaks in the clouds, but just much brighter. Then, just like a missile breaking through the water, we were in the clear, above the clouds and everything was smooth. It was a beautiful sunny afternoon at 6000 feet with only clouds below. It looked as if we were flying over enormous snow sculptures. It was beautiful. I knew then that I had to learn to fly.

Even though some of the training was tough, I was never sorry I did it.

I can't forget young Davey. He's a 14-year-old sailplane pilot who is looking forward with great anticipation to the day when he will be a licensed power airplane pilot. David has wanted to fly for a long time:

I took the controls of an airplane for the first time when I was four and I've been interested ever since. I started working as a volunteer lineman around the airport and took my first lesson at age 12.

I have to admit that sometimes when I'm in school and I should be working on geometry, I'm thinking about how to improve my landings. I love it, especially since I've seen aerobatics in a glider, and I know I must fly. In fact, I've never thought of not flying. I'm totally in control of my life when I'm aloft.

Back down on the earth, I can't even drive a car yet. Flying . . . it's the best time of your life.

Pete told me about his first solo flight.

I was with my flight instructor doing touch and go landings at Bradley Field hoping I'd be ready for solo sometime soon. Air traffic control had us make a full stop landing and the instructor asked me to taxi back to the parking area. I figured we were out of time. When I shut down the engine, she asked me if I had my medical and student pilot license with me. "Yes, I did." My Logbook? "Yep." "Ready to solo?" she said. "Huh?" "Ready to solo?" she repeated. "Yes" I was ready. The big milestone I'd been waiting for since I started my flying lessons. And what an incredible combination of feelings! From the elation of finally being ready to "Well, you're really on your own this time kid," I can think of no other experience that compares to this one, either for sheer joy or just sheer awe. Even though I'm back on the ground, I'm still about fifty feet in the air.

Ever thought about flying around the world? Many people have, but imagine flying nearly 24,000 miles in a light plane. That's what pilot Chuck Classen, an engineer, did with his friend Phil Greth in Phil's single-engine Beechcraft Bonanza. Chuck took his first plane flight as a boy in 1946. Classen watched with awe in 1947 when a local pilot made a non-stop flight from Honolulu to Teterboro, New Jersey.

"I think that's when I knew I wanted to try some sort of long flight myself," Chuck said. "I got my chance in 1988. The trip was an incredible experience. I think one of the greatest sights was, after flying 15 or 16 hours, seeing the Hawaiian Islands for the first time. Something came over me. It was beautiful when we saw the islands over 100 miles out."

Classen's flight set two world records and broke the world record for a single-engine plane (in that weight class) westbound around the world. The two pilots paid the entire $45,000 bill for the trip out of their own pockets, but Classen said "this adventure was worth it." When asked if he could share any advice with someone thinking about learning to fly, Classen said: "If you have any idea at all that you want to fly, just dive in, and try it. If you don't like it, you can always say no later, but at least you tried."

If there is an ultimate flight for a pilot, it would certainly be into space. A man who knows about flying light aircraft as well as spacecraft is former Apollo 13 Mission Commander, James Lovell (Fig. 1-15). Lovell is retired from NASA and the navy, but actively flies his Beechcraft Baron for business and pleasure.

"To me flying is a real sense of accomplishment. I've never stopped loving to fly, even when I was in space. You can talk all day about flying, but if you really want to feel what it's like, you've got to get up and do it . . . go fly. Remember, it takes a while for *anyone* to get used to the flying environment, so be patient with yourself. I only had one ride before I learned to fly. It was enough to make me know I wanted to go on with my flight training."

Fig. 1-15. Former Apollo 13 Mission Commander Jim Lovell talks about space flight during a Palwaukee Airport Pilots Association meeting.

Fig. 1-16. Writer Gordon Baxter.

I asked everyone the same questions and the results varied greatly. The last person I interviewed was writer Gordon Baxter. Bax, as everyone calls him (Fig. 1-16), relates many of his own flying adventures as well as those of other pilots in his monthly column for *Flying* Magazine.

RM: Bax, What made you decide to learn to fly?

GB: To meet more girls!

RM: How old were you when you began your flying lessons?

GB: 34. I was late finding out about girls, too.

RM: Were you ever anxious about learning to fly?

GB: No, just airsick. I always knew I'd be a great pilot.

RM: Can you explain what's so great about being a licensed pilot?

GB: To me, more than anything else, flying is an experience in beauty . . . just like, well, you know.

RM: What has been your most exciting moment at the controls of an airplane?

GB: Without a doubt, any of the 500 or more hours I've spent at the controls of an old Stearman biplane.

RM: Do you have any great stories about your experiences learning to fly that I can pass on to new pilots?

GB: Yes, but I already wrote them all for Flying *and republished them in* Bax Seat *and* More Bax Seat, *currently available from TAB Books.*

Martin Caidin

Fig. 1-17. Some aircraft carry more passengers than others.

2

Learning to fly

NOW THAT YOU'RE GOING to learn to fly, it's a good time to introduce you to some of the terms you'll soon hear people using. First, let's discuss the various levels of pilot licenses available.

The entry level for all pilots, whether they fly airplanes, helicopters, sailplanes, or hot air balloons, is the student pilot certificate. While you hold a student pilot certificate, you may fly only when a certified FAA flight instructor (CFI) is on board. However, in most cases, with a CFI endorsement, you can fly solo for the practice required for your license. You may not, under any circumstances, as a student, carry passengers. When you pass the required exams, you'll be issued a private pilot license. This allows you to carry passengers in a single-engine airplane in good weather conditions (Fig. 2-1).

Glenn Chatfield

Fig. 2-1. Beechcraft Skipper, an ideal primary trainer.

While a private license allows you to share the expenses with someone else on a trip, under the Federal Aviation Regulations (FARs) as the pilot you may not be paid to fly. The commercial pilot license allows you to earn income for your flying. It's not necessary to have a job in mind to earn this rating since many pilots obtain a commercial to learn more about flying.

Should you complete your commercial license and find your thirst for knowledge unquenched, then the airline transport pilot license—the Ph.D. of flying—should make you happy.

In May 1989, the FAA approved the new recreational pilot's license to stimulate flying among people who might not have as much of a business need for flying. This license allows only daylight flight with one passenger within 50 miles of the home base. A pilot's license currently does not carry an expiration date.

CATEGORY, CLASS AND TYPE

The basic private pilot license allows you to fly an airplane (this is the category), single-engine land (this is the class). Type ratings only come into effect if you are flying aircraft that weigh more than 12,500 pounds. Other possibilities are listed below:

Category and class

- Airplane, single-engine land
- Airplane, multiengine land
- Airplane, single-engine sea
- Airplane, multiengine sea
- Rotorcraft, helicopter
- Rotorcraft, gyroplane
- Glider
- Lighter-than-air, airship
- Lighter-than-air, free balloon

Once you understand category and class (Fig. 2-2), it is still possible to add additional "ratings" to any license. The instrument rating allows you to fly a properly equipped airplane in weather that is less than standard visual flight rules (VFR). Weather with a 1000-foot ceiling and visibility less than 3 miles is instrument conditions (IFR).

Another rating to add could be the multiengine rating. This allows you to fly aircraft with more than one engine, something that many people find comforting if they do a great deal of over water, night, or mountain flying.

If you find yourself in need of more knowledge about flying, or are thinking of making flying more than a subtle pastime, the certified flight instructor (CFI) rating would be appropriate. This rating requires a prerequisite commercial license and an instrument rating. The CFI can be broken down into separate ratings of airplane, instrument, or multiengine flight instructor.

Fig. 2-2. Cessna 402, a multiengine land airplane.

MEDICAL

Before being allowed to fly solo in any type of aircraft, the pilot must pass a medical exam. Medicals fall into three classes, third being the least stringent and the first being the toughest. To qualify for a private pilot license, you must pass the third class physical, which merely refers to its certification, not its quality. Don't worry, it's not too tough and it's valid for 24 months. Your flight school or instructor can give you the names of local physicians designated to give FAA physicals.

You make an appointment for the exam that costs about $65. Every two years you must renew your medical. The medical involves a check of your blood pressure, your eyes, ears, and overall health.

If you were to obtain a commercial pilot or flight instructor certificate, you'd need the tougher second class medical which is good for 12 months. An airline transport pilot certificate requires a first class medical which must be renewed every six months.

GROUND SCHOOL

A portion of the road toward your private pilot license is paved with books because you must pass a written exam to win that license. This exam is administered by an FAA-designated examiner in various locations around the country. There may even be a written test examiner on the staff of the flight school you choose.

The written exam covers the following subjects: sectional navigation charts, communications, use of the Airport/Facility Directory and the Airman's Information Manual, aero medical facts, meteorology, weather maps, time conversions, basic and radio navigation, use of the flight computer, aircraft engines, aerodynamics, aircraft performance charts, aircraft instruments, weight and balance, and the Federal Aviation Regulations.

While this may at first glance sound like quite a mouthful, there are some relatively painless ways to pick up the knowledge you need to successfully pass the exam. One method is the local ground school classes run at the flight school you choose. There will usually be a sign-up sheet hanging around your flight school's office asking new students to sign up. All subjects are covered and the classes usually meet one night a week for about 10 weeks. Very often, there are local high school continuing education courses that provide the same service.

There is one problem with this particular kind of training: It takes so long to get through the information that by the time you reach week nine, you've sometimes forgotten what you learned in week two. Luckily, if you are willing to spend a bit more cash, you can arrange to attend one of the weekend ground schools. Illinois-based Flight Standards offers some. Others are offered through the Aircraft Owners and Pilots Association (AOPA).

Flight Standards owner, E. Allan Englehart, mentions a few points for you to consider about the 10-week ground schools. "The instructor of one of these schools may not be terribly experienced. In fact, these instructors may not even be pilots. At the airport schools, the instructor might be a flight instructor from the school who has been told to teach the class and so brings very little enthusiasm. These instructors tend to fill up the weeks with plenty of flying stories to kill time. A new student must ask people who have taken the course, then decide if it's worth taking." The cost for these courses will vary from about $75 in the continuing education version to upwards of $200 for the weekend schools (Fig. 2-3).

The newest method to prepare for the written is to buy the entire private pilot course on video cassette. This method is slightly more expensive than the weekend schools, about $235. It does allow you to work at your own pace and provides the opportunity to review a difficult subject as many times as you need to until it sinks in. Very often, two students buy a set of these tapes together and share the cost (check with each company on this). This can bring the price down to close to that of the longer classroom courses.

Also available now are FAA written exam quizzes and preparation testing for use on your home or office personal computer (see Chapter 7).

The choice is yours. If you're on a tight budget, perhaps the less expensive school is for you. A businessman on a tight schedule, however, will appreciate being able to have the written completed over the weekend. Whether you use the one-night-a-week ground school, or become a weekend wonder, you'll find a method to pass your written exam.

Along with the written exam preparation, each flight lesson should contain some ground school time. This is when your instructor will cover information you'll need to know for that particular lesson. Ground school before flight and the formal classroom (to prepare for the written) should expand your aviation knowledge to the point where you'll feel confident about taking the required oral portion of the flight test.

Fig. 2-3. E. Allan Englehardt teaches students everything they need to pass private and instrument written exams.

THE RECREATIONAL PILOT CERTIFICATE

A new pilot license entered the aviation scene in 1989 and might prove a boon to the aviation community. The recreational pilot's certificate will give more people the opportunity to obtain a license with less training.

The recreational pilot certificate was produced because a committee of government and industry officials believed current private pilot requirements were unnecessarily restrictive to people who only wanted to fly homebuilt and small basic aircraft. With that idea in mind, the new certificate does not carry the requirement for a pilot to obtain specific training in radio navigation, basic attitude instrument flying or radio procedures.

The recreational pilot may fly only small four-place aircraft and may not carry more than one passenger at a time. All flights are limited to within 50 miles of the airport at which flight instruction was obtained. The pilot may not operate from a tower controlled field or into any ATC controlled airspace. Flights are also limited to daylight hours with visibility no less than three miles and not above an altitude of 2000 feet above the ground.

If at some later date the recreational pilot wishes to remove these restrictions, it may be done by completing the appropriate instruction and passing a private pilot checkride.

WHAT DO YOU WANT TO FLY?

Sometimes it's almost like being a kid in a candy store. Airplanes, sailplanes, helicopters, hot air balloons, gyroplanes. Which one do you choose?

The most important idea to have firmly in your mind is how you intend to use your license once you've obtained it. If you plan to use the license to fly recreationally, then an airplane, sailplane, hot-air balloon, helicopter, or gyroplane (Fig. 2-4) might be sufficient. If, however, you plan to be traveling, be it a weekend vacation or from town to town on business, the powered airplane is most likely going to be your best choice.

Fig. 2-4. The Air & Space 18A Gyroplane.

To make an intelligent decision you need some facts about the various types of aircraft and what they can or can't do for you. Most students begin in powered airplanes because single-engine airplanes are more readily available at more locations than any of the other groups of flying machines. They're usually two or four passenger aircraft containing all the radios and instruments necessary for safe and efficient cross-country travel. Single-engine airplanes are one of the least expensive aircraft to fly per hour and a type of aircraft that the pilot can get into the air and fly alone.

A sailplane is considerably less complicated than a powered airplane. It handles very much like a conventional airplane, but lacks an engine and must be towed into the air via a cable, attached to another aircraft. While a sailplane is relatively inexpensive to fly, the cost of the tow plane per flight must also be included.

Hot air balloons are a joy to fly as you stand in the basket beneath the balloon and feel the air rush against your face. The rental cost of a balloon, however, is relatively high, about $100 per hour. A balloon launch requires a ground crew of several people to set it up, inflate, and launch it, as well as time for the crew to follow along behind the balloon to help disassemble it when the landing is complete. Although it's a beautiful ride, it's not for the person who is counting the pennies to their license. It can, however, take much less actual flying time to pick up a private balloon license than a private pilot airplane certificate.

Last is the helicopter. Many pilots in the military have never set foot in anything other than a "chopper" from first flight to their license. One of the primary advantages of helicopters is that they require very little space for takeoff and landing. They're also able to fly at very slow speeds, like 10 or 20 mph, and quickly return to airplane cruise speeds of 150 mph.

One 'copter pilot I know has a helipad (a large square flat surface for the helicopter) behind his house in a Chicago suburb. He flies everywhere, even to the airport to pick up his fixed-wing machine for longer flights. As you might expect, these benefits don't come without a price, which, for a helicopter, is pretty stiff, about $175 to $200 per hour. A helicopter is also more difficult to fly than an airplane. Most instructors I spoke with said the best plan was to obtain a private pilot license in an airplane first. Then you can proceed with an "add-on" rating for a helicopter.

FAA REQUIREMENTS

Any discussion of requirements for your license must be prefaced with information about the government agency we all deal with in flying, the Federal Aviation Administration (FAA). This is an arm of the U.S. Department of Transportation.

The FAA decides what kind of training you must have and what certification standards you must meet to become a pilot. The FAA also writes the FARs, or Federal Aviation Regulations, and is responsible for the nation's air traffic control system. As well as being the people who write the rules, the FAA employs a staff of enforcement inspectors. In locations where there are not enough FAA people to go around, the agency appoints qualified people to perform some jobs, like a designated flight examiner who will probably administer the private pilot test.

A private pilot license, airplane single-engine land, requires a minimum of 40 flying hours logged (35 hours at FAA Part 141 Schools). The term logged comes from the small ledger-type logbook you'll begin and keep for the remainder of your flying career. It shows the date of each flight you've made, the type of aircraft you flew, where you went and how long you were aloft. In Chapter 7 we'll talk about the newest electronic logbooks that run on a home based personal computer and take much of the hard work out of keeping track of your flying time. Of the 40 required hours, 20 must be dual (time when you share the cockpit with a CFI) and 20 must be solo (solo refers to the time when your knowledge and skill are put to the test as you begin piloting the machine alone).

Keep in mind that the 40-hour requirement is a minimum, but it's hardly the norm. In fact, I've never personally known anyone who received a license in the minimum time. To complete your training in that kind of time, you'd need to be enrolled in a full-time school with very little else on your mind but flying. It's something that very few of us can afford.

A more realistic time to plan for your training would be 50 to 60 hours. Recent FAA figures indicate the average private pilot picks up the license at 66 hours. During your flying hours, you'll be learning the basics of how an airplane flies, how it lands and takes off, and the various flight maneuvers necessary to pass the flight test. Another portion of the training involves use of the radio and navigational aids on cross-country flights to teach what is necessary to safely complete long distance flights. Additionally, you'll be spending a good deal of time learning the specifics of the mechanical systems that keep an airplane flying.

Within that 50 to 60 hours of flight time is at least three hours of dual instruction in cross-country flying. This is designed to sharpen your skills so you can safely complete the 10 hours of solo cross-country flying necessary for the license.

You'll also have three hours of night flying with your instructor. If for some reason you are unable to fly at night, you can receive a restricted license that permits daytime visual flying only.

There will also be at least three hours of dual instruction in direct preparation for the flight test.

The requirements for a helicopter are very similar to an airplane in that they also require 40 total flying hours minimum with the same sort of ground school subjects needed for an airplane (Fig. 2-5).

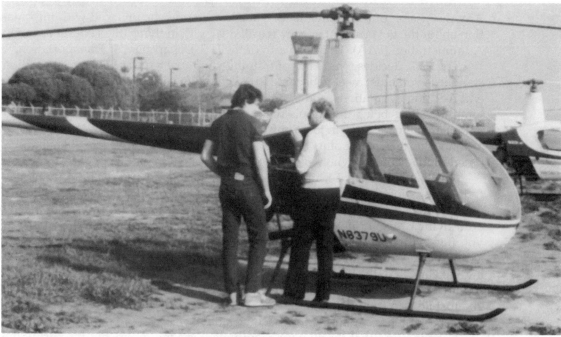

Fig. 2-5. Student and instructor discuss an upcoming lesson.

A sailplane license does not allow the flexibility of a powered aircraft license. You can remain aloft only as long as warm air thermal action provides the lifting components the glider requires. However, there are some pluses to the sailplane rating.

With a helicopter, airplane, or gyrocopter private pilot certificate, the applicant must be 17 years old. A sailplane or hot air balloon pilot may be licensed at 16. A student for a sailplane or hot air balloon need be only 14 years old to begin.

Another important factor to consider is that the total aeronautical experience for the hot air balloon and sailplane license is much less than a powered airplane. The FAA requires a total flying time of only 10 hours before you're eligible for the hot air balloon license. The average is a little higher, about 15.

A private pilot-sailplane certificate can be issued with as little as 7 hours of solo flight including 35 glider flights launched by ground tows or only 20 flights launched by aero tows (Fig. 2-6).

Fig. 2-6. Sailplane on a tow line to altitude.

Most of the various pilot certificates after the private also carry flight time requirements before you can qualify for the FAA checkride. The commercial certificate demands 250 hours of total time while the instrument rating requires at least 125 hours of logged time before the checkride. Many flight schools will combine the commercial and instrument rating into one program. The multiengine rating carries no specific hourly requirement, nor does it require a written exam. You train only until you can perform the specific tasks required for the rating. If you train for the airline transport pilot certificate, you'll need at least 1500 hours logged before the exam. Before that exam, you'll need to have your logbook certified by a local FAA office, so be sure you've kept accurate records.

If your goal is to pick up the quickest private pilot license that you can, it's probably going to be the hot air balloon or the sailplane. However, if you're looking for the most useful license, the one that will allow you to travel quickly from Tampa to Miami and back in a weekend for a visit with friends, the private pilot airplane category is going to be the best bet. It's a good balance of all the possible combinations.

Flying a powered airplane gives you more flexibility than most of the other flying machines and while it does cost more than the sailplane or hot air balloon, it is considerably less than the dollars you'd probably spend to get a private pilot helicopter license.

By this time you've probably decided just what you'd like to fly. If you're in powered airplanes or gliders or helicopters, there are several models to choose from.

In powered airplanes, two common models are the Cessna 152 or a Piper Tomahawk. They are usually the cheapest to fly, too. If you find strapping on one of these is a bit too tight a fit, try a Cessna 172 or a Piper Warrior or a Beech Sundowner. All of these models normally seat four people and give you more interior room as well as excellent training in a good cross-country airplane. We'll be discussing more about specific models in later chapters.

If helicopters are your interest, you have fewer to choose from. Some of the common training machines are the Enstrom or the Robinson. Leave the big 'copters until after you have your rating. (That pretty much holds true with whatever you begin flying.)

If you're not sure which model you want to start in, try a few different ones until you connect with one you like. The less complicated the aircraft is, the easier it will be for you to learn. Remember, it's your money. No matter what, fly what you like.

A VISIT TO THE AIRPORT

If you're considering learning to fly, you've probably already visited your local airport. If you haven't, my instincts tell me you know where it is. Let's spend a few moments and visit a local small airplane airport and give you an idea of some of the things you'll see and some of the terms you'll hear.

First, if you're going to be learning to fly at an airport that caters to small aircraft, you should understand the difference between small airplanes and big ones and the way most people in the business use these terms. There are three groups of aircraft: air carrier, military, and general aviation.

Air carrier includes everything that flies regularly scheduled routes in aircraft that usually weigh more than 12,500 pounds; Boeing 727s, 747s, McDonnell Douglas DC-9s, and the like, fall into this category. Next we have the military, which includes any airplane or helicopter from any branch of the armed forces.

All the other aircraft are general aviation, but there can be quite a wide variety of types and sizes that fit into the general aviation group. They range from trainers like the Cessna 150 or Cessna 152 or Piper Warrior that weigh in at about 2000 pounds, and fly about 120 mph, to larger twin-engine propeller-driven aircraft, like a Cessna 310 or a Beechcraft Baron, usually used by companies for transportation of their executives and customers. The top end of general aviation are where you will find the large corporate jets.

General aviation has some of the most sophisticated business-owned jet aircraft around. In fact, until the new generation of air carrier aircraft, such as the Boeing 757 and 767 appeared in the mid 1980s, general aviation was the place where the sophisticated electronic navigational gear was being used. One corporate-owned aircraft is the Gulfstream IV (Fig. 2-7). It can carry 14 corporate executives in comfort and splendor from New York to London non-stop and is worth up to $20 million. The corporate side of general aviation is usually referred to as business aviation.

Gulfstream Aerospace

Fig. 2-7. Queen of the corporate fleet, a Gulfstream IV.

Figure 2-8 is a typical general aviation airport layout. It might have only one runway or as many as three or four. The runway is usually hard surfaced or it could be nothing more than hard rolled grass outlined with white tires. The length of the runway is determined by the kind of aircraft using it regularly; larger aircraft typically need more runway than smaller ones.

The choice of hard surface or grass surface also indicates the aircraft that will arrive. Grass runways are used primarily by single-engine and some light twin-engine aircraft typically weighing less than 6000 pounds. Much heavier than that and they'd probably sink into the sod. By comparison, a corporate jet like a G-IV weighs in at about 65,000 pounds.

Usually, each runway will have a taxiway of some sort to allow the aircraft to clear the runways and move to and from the arrival ramps and hangars. Most airports have at least one large ramp area. These service ramps look much like a large automobile gas station. The entrance allows aircraft to taxi right up to the pumps to have fuel tanks filled and the oil and tires checked.

Located near the service area is the flight office. This is the main office for the fixed base operator (FBO) and includes fuel sales, aircraft maintenance, flight

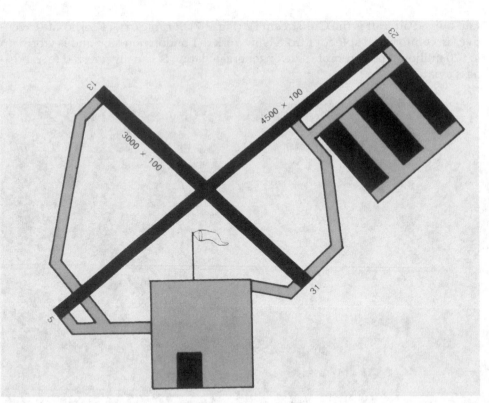

Fig. 2-8. Layout of a typical general aviation airport.

training, flight charters, and, sometimes, aircraft sales (Fig. 2-9). Depending on the size of the airport, there may be just one or as many as four FBOs on the field, all competing with one another. This competition is usually good for the customer.

Fig. 2-9. Fixed base operator (FBO) maintenance hangar.

Each FBO usually has at least one large hangar to provide the maintenance necessary on both an airframe and its engine. Services range from changing a tire to a complete overhaul of an engine. All maintenance on aircraft is performed by FAA-licensed mechanics. Our sample airport also has a radio shop to keep all the aviation electronics, commonly called avionics, operating at peak efficiency with the help of an FAA-certified radio technician.

In various locations around the airport, you'll find small hangars with doors about 10 feet high. These are called T-hangars because the aircraft are parked next to one another (forming Ts) and are used for inside storage of single- and twin-engine airplanes.

Other aircraft owners will usually keep their airplanes tied down in a grassy area near the ramp at a rental rate considerably less than the rental cost of a hangar.

Visible during the day alongside the runway near the portion where the aircraft begin their takeoff roll, you'll find a large international orange-colored piece of cloth that resembles a large sock (Fig. 2-10). This is, in fact, called a wind sock and helps a pilot determine from which direction the wind is blowing. Aircraft always take off into the wind to shorten their ground roll, so knowing the direction the wind is blowing is always of utmost importance. Through communication on a specified radio channel, most pilots can determine the wind, pick out a runway, and land on their own at airports that don't have a control tower (Fig. 2-11).

Fig. 2-10. Runway wind sock.

Along either side of the runway you'll see lights set up on small pipes about 10 inches above the ground. These are the runway lights that help define the landing area for pilots at night. Runway lights are always white. As you exit the runway at night, you'll see hundreds of blue lights in a seemingly endless collage that outline the taxiways. As day blends into night, red lights will appear scattered around the airport on top of objects within a few miles of the airport. These obstruction lights help the pilot avoid antennas, water towers, and the like.

At some airports you'll find a tall building with plenty of glass and radio antennas on top. This is the control tower. Inside, highly skilled air traffic controllers use radios and binoculars to sort out in-coming and out-going flights. They keep

Fig. 2-11. The Sedona, Arizona, airport serves many different types of aircraft.

things running harmoniously on the airport. Most control towers are run by the FAA; the controllers are employees of the U.S. government.

A control tower usually provides one of the most panoramic views of the airport you'll find. Try calling them the day before you plan to visit the airport and they might arrange a tour of the facility. Plan to go back for another visit after you start training.

HOW MUCH WILL IT COST?

Nothing seems to change faster in America than prices. With that in mind, you'll understand why the rates I give you in my examples could well be out of date by the time you read this book. Except for the mid-seventies, we have not had much more than three to four percent inflation per year. It has been a creeping inflation, just enough to remind us pilots that the cost of flying might go up.

There are two different operations you can turn to for a private pilot airplane license. One is the FBO at your local airport and the other is the freelance instructor. Each has advantages.

The FBO will usually, although not always, have the most up-to-date fully equipped aircraft as well as the latest in video ground school equipment to help you through the hurdles of learning. Whenever a problem occurs with the aircraft, it's a pretty simple task to have the airplane pulled into the FBO's maintenance hangar and let one of the FBO's mechanics work on it. The problem for you, the consumer, is that all this equipment and personnel are overhead and their cost must be reflected in the price you pay for those flying lessons. (An increase in the property taxes paid by the airport alone can also play havoc with prices.)

The person who isn't as susceptible to these problems is the freelance instructor. The freelancer often owns just one airplane somewhere on the airport. The preflight and postflight ground school is usually conducted with a legal pad and pencil right under the wing of the aircraft or in the front seat of a car before you fly. One instructor and one airplane doesn't cost nearly as much as the lease of buildings, a half-dozen aircraft, 10 or 15 employees and all the equipment to keep everything flying. This can mean dollars saved. An FBO might charge $50 an hour for a Cessna 150 plus $23 an hour for the instructor. The freelancer might be able to give you a slightly older 150 for perhaps $40 per hour plus $18 per hour for the instructor. Let's see just how much difference that can make to you.

FBO rates—minimum

30 hours dual at $73 per hour = $2190

25 hours solo at $50 per hour = $1250

25 hours ground school at
$23 per hour   = $ 575

Total $4015

Freelancer rates

30 hours dual at $58 per hour = $1650

25 hours solo at $37 per hour = $ 925

25 hours ground school at
$18 per hour   = $ 450

Total $3025

That $1000 difference is like getting almost 27 hours of free flying from the money you save working with the freelancer. At first glance you'd have to ask why everyone doesn't run out and find a freelance instructor and gain their license that way.

The answer to that is the same as why some people will leave their savings in a passbook at 3 percent and others will put it in a money market fund at 5 percent. Some people are bargain hunters and are willing to do with a bit less (in savings, it's security) to gain more in the end. In flying, some students feel they must have the best of everything because anything less is unsafe. There are even FBOs that will tell you that freelance instruction is not as good as that from an FBO and that might not be true.

Certainly you should understand that an FBO or a freelance instructor is a businessman like anyone else. Their goal is to make a living off something they enjoy doing, in this case teaching people how to fly. I've known some pretty poor instructors who worked with some terrible equipment at FBOs. I've also know some freelancers who rented top notch equipment and were excellent flight instructors.

One item sometimes used against freelancers is their lack of organization. FBO flight schools are normally operated under the scrutiny of FAR Part 141. The FAA visits the school on a regular basis to make certain the FAA-approved curriculum is being followed. A freelancer may work under the less stringent FAR Part 61 rules and never meet an FAA inspector except every two years for the formal renewal of their flight instructor certificate.

The Part 141 school must be better organized and should be capable of providing better training. This is not necessarily the case, however. Part 141 organizes training into subjects that must be followed in a certain order. It also describes what must be taught before the student is eligible for a license. This is in addition to the basic regulations for licensing that we all follow under Part 61. So what does all this mean to you the student? In reality, it means extra paperwork and more cross-checks on your progress. If the school in question were to lightly skip over some areas and just "paperwork certify" you, your education would probably not be any better than what you'd get from the freelancer. Part 141 schools have been praised because under that part you can get your license in 35 hours. Most of the time though, no one can learn all the information that fast, so it really all becomes academic. Some Part 141 schools can reduce the required time for a commercial instrument rating down to 190 hours total flying time too.

I began my instructing career as a freelancer and had students singing my accolades after the first group of them picked up their licenses. This wasn't because I had a better set of rules to follow. Rather, I taught a great deal of hands-on subjects such as what to do if the engine quits or what the best way is to handle a busy tower frequency when you're a student. My students came away with a very practical grasp on flying around a big city.

I have been on both sides. As a flight department manager for an FBO, I worked closely with the Part 141 program. I honestly don't think it can be said that a freelancer is a poor instructor and a Part 141 teacher is good. It really depends on the instructor and the company, but you must remember, you're still a consumer and you must do your homework.

Finally, consider how you will pay for your training. You should be trying to fly twice each week from the start. Sure there will be times when you get rained out or you're sick, but twice a week should be your goal. At that rate, you'll retain the training information better and you'll see good progress. However, if you fly only when you have the money, it could take forever. I recommended one of my students take a loan to pay for his flying. Besides being able to absolutely fly every few days, he earned a better block rate on the price of the aircraft. If you are working toward the more professional pilot certificates and you served in the U.S. Armed Forces, check with your local Veteran's Administration office to see if you qualify for the New G.I. Bill. If you do, you could find Uncle Sam picking up 60% of the cost of your flying.

CHOOSING THE RIGHT INSTRUCTOR

How do you know the instructor you choose will do the best job for you? The answer is that you don't, at least not at first. Learning to fly is a two-step process: first you get the learning environment in order, then you make certain that the teacher has the right attitude. Choosing is really not too difficult, once you learn what to look for in an instructor.

Finding the right school and instructor can be as easy as looking in the local yellow pages under aircraft training. If you live in a large metropolitan area, you'll have more choices than in a rural community. If you want to try a freelance instructor, you might just search the local newspapers under the aviation or recreational want ads. A freelancer's ad might be simple: "Learn to fly in my airplane. Great rates, call Jim."

Whether you begin with the FBO lists in the phone book or the freelancer's ads, keep a legal pad at your desk when you begin making phone calls. Ask for the chief flight instructor (or the only flight instructor, if it's a freelancer) and say you're considering their school for training. Instead of letting that person tell you what they think is important, try asking questions: How many full-time flight instructors do they have? (More full-time flight instructors means easier scheduling for you.) How many aircraft does the school own and how old are they? (If you can fly a newer airplane for 5% more than a 10-year-old model, take the newer one. It's probably better equipped.) How long it should take, on the average, to complete your training. If you talk to a freelancer, ask how often the plane is booked on the days you would normally want to fly. Also, in both cases, ask for the names of some recent graduates as references. Must you pay at the end of each lesson or will they bill once a month? What is the school's policy on refunds or being late to a lesson or even canceling a session?

Now, it's time to visit some of the airports to match what you've heard with what they really have. Now is the time to let the instructors "sell you a bit" on why you should choose their school. Did the boss spend some time with you or did he push you off on another instructor? "Hey Fred, here's a new student for you." That attitude probably indicates how they'll handle other problems, too. How do you know whether the attitude is because the instructional staff is too busy or just doesn't care? The answer is you don't, so perhaps this school isn't for you.

At one very large school I visited before I began my instrument training, I was shown into the school founder's office. I was given a cup of coffee and made to feel they cared about the money I was thinking of spending. I liked that feeling. The school was well worth the money I spent there.

So how do you know when you've found a good instructor and a good school? A gut feeling is probably a good bet, but there is nothing like getting into the air with an instructor to see how that person reacts in the working environment. Most schools offer a no obligation introductory lesson for $25, or perhaps $50 if it's a helicopter school. It's your chance to check the place out. If you like the way things go, sign up.

A special time for student pilots is the solo, that first time aloft alone. There is nothing quite like watching the ground fall away as you climb to altitude in an airplane all by yourself, knowing your instructor has confidence in your abilities.

Think for just a moment before you go off on your own. How much confidence do you have in yourself? You've got to believe you can do it. How much about flying have you learned up to this time? Remember, the idea here is not to do the instructor's job. Be an active participant in what is happening and offer an opinion on a situation if you think it's called for.

A solo at seven hours may surely impress people on the ground, but did your instructor cover emergencies or crosswind landings before he signed off your ticket for solo? How about how to handle people cutting you out of the traffic pattern? If your instructor never mentioned any of these things before he let you go, subtract more points and find a new instructor.

Anyone can handle an airplane when everything is going perfectly. When the going gets rough, you'd better be prepared. You can't be if your instructor was more concerned about how soon he got you up flying by yourself than he was about how safely you got up there.

Think of your solo as a flight instructor does. It's a confidence building maneuver. It's designed to demonstrate that yes, you do have what it takes to learn to fly but you still have a way to go. Believe me, it's better to solo at 12 hours and have already performed emergency landings, with some degree of skill, and been introduced to crosswind landings, with an instructor in the right seat to coach you along, than to find yourself panicking in the traffic pattern if the engine coughs or the wind shifts. Learning to fly isn't a contest. More hours before solo doesn't mean you're stupid. Waiting for the solo until both you and your instructor feel truly confident shows the kind of maturity you need to become a competent pilot.

By the time you've completed the first five hours of flight instruction you should understand the purpose of every dial and switch on the instrument panel. If you don't, if your instructor is the kind who says "Put the airspeed at 70" or, "This lever works the flaps," get a new teacher. Obviously, this person is probably going to skip a lot of other important things, too.

An old flying cliché says that the runway behind you on takeoff is useless. So is the knowledge in someone else's head. No instructor can teach you everything, but certainly can't answer the questions you haven't asked. It sounds trite, but there really are no stupid questions, even if it's the fifth time you had to ask. Remember, that instructor is working for you, not the other way around.

Be wary of the conflict between ground time and flight time. In some operations, an instructor makes more money in the air than on the ground. This often means the instructor will spend five minutes on preflight ground instruction and the rest of the hour trying to teach in the airplane. An airplane makes a poor classroom because it's noisy, sometimes hot, and it requires the student and instructor to maintain a constant vigil outside for traffic.

If the instructor is initially trying to explain in the airplane how to perform stalls or how the oil pressure gauge works, get a new instructor. The philosophy of flight should be discussed on the ground in a quiet atmosphere, and illustrated in the air. An instructor can talk and fly, but to expect you as a student to be able to grasp complex concepts while you're trying to fly is unacceptable.

Sometimes the pressure to fly, instead of talk on the ground, can cause instructors to try things they might not normally do, to make more money. One instructor I know took out a new Cessna 172 for touch-and-go landing practice in wind of 30 knots gusting 42 knots. While the actual landings and takeoffs passed without a hitch, the downwind turn off the runway towards the ramp ended with the Cessna flipped over by a wind gust.

The instructor was fired for not exercising better judgment, the airplane was wrecked, and the student's confidence bruised. A chat with your instructor about limits might be very enlightening. If you don't feel comfortable with your instructor's limits, ask for a new teacher.

You've flown with your new instructor for a while now. How do you feel? Are you really learning something or are you leaving each lesson with many unanswered questions because the instructor either doesn't explain fully or doesn't have time? If you're not getting what you want, buy the instructor a cup of coffee and discuss where you feel you stand in your training. It might help. If the instructor says he doesn't understand or doesn't seem to care, find a new teacher.

How did your instructor do? Did he or she seem genuinely interested in your progress? Did they quiz you often to test your absorption of the material? Did they answer your questions fully? If the answer to all these questions is yes, this instructor is worth his or her weight in gold because they're teaching you not only how to fly, but how to think and act for yourself—like a pilot—and that's really the point of it all.

I've discussed what obtaining a private pilot license entails in money, time, and mental resources. I've given you some tips on the best way to go about selecting the school and airplane you'll want to use. So what's stopping you from taking that first lesson? The answer? Nothing. Absolutely nothing.

3

Student ups and downs

I WISH I HAD A DIME for every person I've talked to about learning to fly who looks interested until they realize there's classroom work and studying involved. They roll their eyes up and the expression on their face looks like someone who just swallowed a piece of lemon. I'm not going to sit back and tell you flying an airplane is as easy as driving a car, because it's not. But it isn't as tough as studying for the bar exam either.

Even though more and more people graduate from high school and college now than ever before, they often leave school with an intense love of the social activities and a distinct hate for studying. If you know there is no way in the world you are ever going to crack a book again, no matter how enticing flying might look, then stop reading right now because you'll never become a pilot, never. If you're able to put some of the old devils about studying to bed and try some new ideas for studying that have popped up in the aviation spectrum the past few years, you'll find the reward well worth the work.

The nice thing about learning to fly is that you work at your own pace. There is no one to stand behind you and tell you to do your homework. You work on your own. What you do have in flying is a very dramatic display if you don't keep up with the book work. For instance, your next lesson might be about slow flight, how to enter it, how to control the aircraft while you're at a critically slow airspeed, how to return to normal flight, and various pitfalls. You might walk out for your lesson and think you can do it without studying. When you get in that airplane and start trying to perform and realize that the airplane is very quickly slipping away from you, you'll understand the importance of the theory.

If that kind of embarrassment doesn't affect you, then let me tell you now that the harder you fight against the book work portion of flying, the more money it will cost you to complete your training. If the instructor must force-feed you everything you need to know, you'll find the entire training process slowing down, too. You won't complete an entire lesson plan each session because you were not prepared. If you want to make the best use of your money, the instructor's time, and pick up your license in the shortest amount of time, sit down and look this venture right between the eyes. Commit to keeping up with the lesson plans and required studying for each session. In actuality, we are not talking about a great deal of time spent on your classroom studies either. Spend an hour a night, twice a week studying the material and forming the questions to ask your instructor at the beginning of the next lesson. You'll find yourself moving along very quickly to your goal of becoming a private pilot.

SCARY PARTS

I use this term loosely and almost humorously because the stories some students bring to their first lessons are often enough to boggle the mind of most instructors. Consider a student I had who came out for his introductory flight but wanted to cancel once he arrived. He'd heard about stalls and decided, upon seeing the small Cessna 150 we'd fly, that no one was going to take him up there and shut the engine off, even for a demonstration.

When I looked at him in amazement, he explained that his brother-in-law had read a story about someone else who'd learned to fly and could never get past the lesson on stalls. My new student and his brother-in-law had made the same mistake that hundreds of students do each year by trying to relate flying to driving. My student knew that when his car stalled, he was usually forced to pull over to side of the road. He also knew if the engine on our trainer stalled, he'd have no place to park.

Very calmly, I sat down under the wing of the airplane and showed this new student what stalls in an airplane were all about. I explained that stalls in airplanes had nothing to do with the engine of the airplane at all, but rather with the airflow over the wing. I went on to explain just how stalls are performed.

I told him that at no time did I ever turn off the engine. If I, as a student, had an instructor intentionally turn off an engine on a single-engine airplane in flight, I'd get back down on the ground and find a new instructor. There are enough things to learn in flying that can cause students anxiety without actually putting the entire flight in jeopardy.

During your training it is, however, quite common for an instructor to pull the throttle back, the equivalent of taking your foot off the gas in your car, to simulate engine failure. The engine never actually stops running. The maneuver is designed to orient you to the realm of flying and one of the toughest challenges some students must undergo. You must learn how to plan, and plan well, if you're ever going to become an excellent pilot. I think every student should strive to be the best. The world is filled with millions of people doing mediocre jobs and flying is something that reveals a bad pilot pretty easily.

Emergencies do happen in airplanes. Most of the time, though, pilots were given some early signs that were either ignored or unnoticed until the situation became critical. Here are a few examples. A pilot has a great deal of difficulty starting his airplane on two or three occasions. An automobile driver might say, "I'll have to get this thing tuned up one of these days." Remember, though, in an airplane, safety is the most important concept.

What this pilot should have done was to get the airplane over to the maintenance hangar on his field and have a mechanic look at it. Unfortunately, this pilot treated his airplane like his Chevy and kept flying. En route to a weekend adventure, the pilot found the engine began running rough. Now the pilot tried to figure out what the problem was, but before he could, the engine stopped running altogether. The pilot managed to put the airplane down in a farmer's field and luckily no one was hurt. This accident could have been prevented if the pilot had been in tune to his airplane.

Most new students believe that an engine failure is the worst problem that could possibly happen in an airplane. While it's nothing to poke fun at, some of the consequences stemming from an engine failure are not correct. If the engine on an airplane stops running for any reason, the airplane does not come tumbling out of the sky. It becomes a glider, fully controllable as it descends. Sure, the pilot must look for a place to land while he tries to restart the engine, but all the while, the pilot is in complete control of what is happening.

Another student had been criticized about sloppy preflight inspections. Underneath the engine compartment is a small drain designed to remove any moisture that might have condensed in the fuel tank. The student would occasionally not wait long enough to see if the fuel drain was closed before she'd move on to something else. This day, in preparation for a cross-country flight, she checked the fuel but did not wait to make certain the drain was fully closed before moving on in the inspection. On this day, luck was not with our pilot.

Forty minutes into the flight she looked at the gauges and was astonished to see the needles indicating only ¼ full even though she knew she'd started with full tanks. At this point, the student assumed the gauges were broken and proceeded past three more airports before the engine ran out of fuel. The only piece of luck this student found was that the engine quit as she was flying over an old country road where she did manage to glide safely down to a landing. Again, poor preflight and poor judgment caused the accident. The gauges told the truth. The pilot just didn't believe them.

During your training, you'll be introduced to a great many different problem situations. The idea is to teach you, as a student, to plan well and use good judgment so when clues that foretell a problem appear, you deal with them before the problem becomes a crisis.

You're going to learn that when you fly, the weather can change from what you had been told to expect. You will be trained to listen to the weather forecasts and then apply what you've heard to what you see out the cockpit windows as you fly. You'll learn about the problems that weather can cause and how to detect these problems so you can alter your course or turn around before a little problem becomes a big one. No one can teach you judgment, but they can teach you what to do and not do in a certain situation. You learn. When a problem arises, you synthesize solutions from what you do know.

SCHEDULING

A good training environment demands a commitment from the instructor and the student if you're to reach your goal of a private pilot license. An important idea to be considered before you begin to fly is the schedule you'll be able to work with. This not only includes your time frame, but that of the airplane and instructor.

In any learning experience, it's been proven time and time again that repetition is not only important, but one of the critical factors to your success. The more often you're exposed to flying or anything having to do with flying, the more quickly you'll learn. You'll also find that retention of each lesson's material will also be greater. With that in mind, you must be ready to fly at least once a week, minimum. Flying less than once a week is a losing battle, because you'll spend half of the new lesson reviewing the material of the previous lesson.

If you can commit to one lesson per week and stick to that schedule, you'll find yourself making good, observable progress. Some people are going to shake their heads at this schedule by waving their checkbook: "But I can't afford one lesson per week." I'll assume by this time you've already decided you want to fly. Motivation is not the problem. Perhaps your budget is. Did you know that with less than one lesson per week, the total bill for the license will actually be higher than if you'd just moved ahead on schedule and let the other things fall where they may?

It makes sense really, if you think again about the above example of the student who doesn't fly very often. Each week the instructor repeats large portions of previous lessons because the student forgets. This means there is less time to move ahead to new material. It might not sound like much at first, but if you lose half an hour each lesson, in just a few months you'll be four or five lessons behind a normal schedule.

I remember a student who took two and a half years to pick up his private license and it had nothing to do with his intelligence. He started strong and made one lesson per week but slacked off after about two months. He'd miss a week and hadn't flown in two by the time I'd see him again. Before we knew it, we were involved in one of the worst winters in history and we often could not fly because of bad weather. (Remember, you're going to be a fair weather pilot at first.) Three or four weeks would go by in winter before we'd be able to get together in the airplane. By then, we learned almost nothing new, and spent most of the lesson just trying to figure out what he retained from the previous month.

The point is that you're really throwing your money away if you don't make the commitment to fly often. Certainly, if one lesson per week is good, two is better. Schedule one lesson for Saturday and another for Wednesday. If you should miss one for personal or weather reasons, it will be but a few days until the next. You'll make great progress at this rate. The same student in the previous example was always canceling lessons. He was short of cash. While we must all consider our budget in this flying game, this man's checkbook went dry so often that he added even more time to his license and by the end of the process, had spent an extra $1200.

Near the end of the second year, I had a talk with this student and explained what was happening. He already felt frustrated and I was on the verge of asking him to withdraw because his progress was so slow. He told me that he just couldn't seem to budget for his lessons very well. I came up with a solution that made life a great deal easier on his pocketbook as well as more rewarding for the both of us. I advised him to go to his bank and apply for a loan for the remainder of his flying lessons which totaled about $1200. Within just a few weeks, he had opened a separate checking account with the loan proceeds and used that only for flying. From that point on, this student flew twice a week and within three months had taken and passed the flight test.

PROPER ATTITUDE

Whether you choose a freelancer or an FBO as your place to learn, your attitude toward learning is just as important as how much money you have to spend or what kind of airplane you learn in (Fig. 3-1).

Fig. 3-1. A pilot heads out for some solo flight training.

You must believe in yourself. You must believe you possess the talent and ability to learn to fly, or you might as well not waste your money or time. Some people can dive into flying aggressively and cope easily with the problems of learning while others don't fare as well. They bring to flight training some of the same baggage they carried with them in high school and college. The inability to ask questions for fear of appearing stupid or allowing themselves to be totally intimidated by the instructors—even when the instructor is actually trying to get the student to open up—should be left at home. Earning a pilot's license is accomplished one step at a time, one day at a time. Go to the end of this chapter when you've finished reading this book. Make a copy of that license you see there. Heck, cut it out if you must. Then take a little white-out and cover up the name on the license and type in your own. When all this work is completed, put the copy of YOUR license somewhere you'll see it each and every day . . . on the fridge, the mirror in the bathroom, anywhere you'll be able to look at it more than once a day. I guarantee that daily reinforcement of your goal will help see you through to the real copy of your license.

Flying an airplane isn't a passive activity; it requires you to absorb a great deal of information to make good solid decisions on your own. Again, no instructor can teach you good judgment, but a good instructor can teach you how he or she decides what to do. As your aviation knowledge increases, you'll be able to make similar good decisions. This kind of brain training comes in handy when you're presented with a situation you've never encountered before. It's then that you'll have to synthesize a solution to the problem based on your own experiences. Only a student who participates in the training program is going to learn what is necessary.

As a student, you must ask questions all along the way to be certain that you understand what your instructor is showing and why he makes certain decisions. If you don't know the whys, you'll never be able to make good decisions on your own. Flying is no place for someone who needs to be force-fed information or who only wants to know enough to get by. If you don't truly take an active part in your flight training, your chances of ever becoming a good pilot are slim.

FIRST LESSON

There are plenty of exciting segments in your flight training, but one of the most exciting is the first flying lesson. The decisions have all been made and the budget has been manipulated to include enough space for lessons. By now you've managed to convince the skeptics in your family that flying is something you really want to become involved with and today's the first day. The weather will hopefully be sunny with light winds and no matter how much you try, you'll still find yourself totally enamored by everything you'll see in the cockpit. Believe me, everyone that flies has been through this same anxiety.

Be sure you arrive about half an hour before the lesson is due to begin. There is nothing worse than arriving late and feeling rushed. Not today. In our first lesson, the instructor greets you at the door and sits you right down in his office to get the paperwork going. The paperwork includes getting you set up for an FAA physical, a logbook, and a training folder. After that, the instructor checks out the airplane and walks to the parking area with you, pointing out the important landmarks along the way, the control tower, the fuel pumps, the maintenance hangars, the parking ramps, and tie downs.

As you reach the plane, the instructor begins a preflight inspection, explaining the names of the various parts of the aircraft and their purposes. The instructor should touch lightly on the fuel supply, the propeller, and the engine. Enter the aircraft and adjust the seat. Your instructor will give you basic information about the instruments and their purpose, as well as the various switches and levers. Keep in mind that you're not expected to remember all this information the first time.

Rather, try and assimilate it with the reading you'll be doing in the textbooks the school will ask you to purchase.

Once you've watched the engine start up, the instructor might ask you to rest your feet lightly against the rudder pedals so you can feel what's happening as you taxi and perform the pre-takeoff checklists. Checklists, by the way, are something you'll see a great deal in flying. I've had students watch me run through the printed checklist and chuckle, "What's the matter, don't you remember what to do next?" Quite frankly, I do remember what to do next, but the checklist is a backup to make certain. If you are going to be a good conscientious pilot, learn this now . . . good pilots use a checklist. It could save your life someday.

Once you've completed the run-up, the instructor might again ask you to rest your feet lightly against the pedals, being careful not to touch the brakes. The instructor might also tell you to lightly grasp the control wheel so you can feel what's happening during the takeoff roll. You'll be amazed at how quickly the aircraft leaves the runway, but then most trainers weigh only about 1800 to 2000 pounds fully loaded, possibly much less than your automobile.

Now that you're flying, listen as the instructor explains how the aircraft responds to certain movements of the controls. Once you've reached cruising altitude, the instructor will demonstrate some of the basics of flying, such as turns, slow flight, climbs, descents, and, of course, straight-and-level flight. A good instructor is going to let you try a few of these maneuvers on the first flight just so you'll learn how really easy they are to perform. Before you know it, you'll be turning and climbing on your own under the instructor's watchful eye. Too soon, though, the hour's lesson will be finished and you'll be headed back to the airport and landing. Watch and listen, but don't be too concerned if you don't grasp everything that's happening. This is normal. At the completion of your lesson, the instructor will review the lesson and talk about what to expect next time, then assign some reading.

RADIO LINGO

Most of the aircraft you'll learn to fly in have at least one communication and navigation radio installed. According to the FAA regulations, a radio is not necessary to fly an airplane in certain parts of the country away from areas of dense traffic. For most of you, learning to listen and communicate on the radio is a fact of life, however. It's something that even the shy will eventually learn to operate with ease, if they don't let the radio and its language get the best of them.

One of the biggest problems new students seem to face is being "microphone shy." Learning what to say into the microphone is really a simple matter, though. The problem usually arises when the student is actually faced with pushing the microphone button and talking when they know hundreds of people might be listening. I've seen the CEOs of major corporations turn white as they tried to squeeze the life out of the poor microphone, hoping that it and the task of talking into it would disappear.

Learning to talk on an aircraft radio is a little like acting. No one fumbles when you know your lines. To this day though, I've had situations in which I was asked for something on the radio unexpectedly and did a few "ahs" or "ums" before I managed to figure it out. So the main thing to remember is that no one is going to be perfect at radio operations. There is no such thing. Airline pilots fly the same route over the same navigational aids, and land at the same airports, so many times in a week that they sound totally cool all the time. Everybody would sound good with all that rehearsal.

If you know your lines, you can sound like a pro. If you're operating out of a tower-controlled field, your instructor should review radio procedures with you. But some instructors prefer to let a student just dive in and take his chances. I think that's a big anxiety producer, as well as a big waste of time. Your instructor should be able to give you, in two typed pages, the typical conversations between aircraft and ground controller, and between aircraft and tower for landings and takeoffs.

He'll also teach you how to listen to the automatic terminal information service (ATIS) if your airport has one. It's worth your time to sit down with your instructor and make certain you understand what the various radio phrases are and when you'll be using them.

Now comes the easy part for some people. Memorize the phrases. That's right, just plain memorize them. The reason is simple. If you know exactly what you are supposed to say, you only need to learn exactly when to say it. The reason so many people become microphone shy is they never once think about what they are going to say until they press the button on the microphone. That's when their mind goes blank.

The other half of the problem is making sense of all the "lingo" you hear when the radio is on. Again, the solution can be pretty simple. There are only about 10 phrases you'll ever really use in ATC communications and if your instructor was diligent, he already told you what they are. One of the simplest ways to improve your understanding of radio lingo is to review the list of phrases, then go out and buy a battery-powered aircraft band radio. Pick one that has a variable tuning knob. It will be much less expensive than the larger programmable digital units and, besides, you most likely wouldn't know all the local frequencies to tune anyway.

Take your receiver out to the airport. Put yourself in a position where you can watch part of the traffic pattern as well as the last half mile of the final approach. Turn it on and listen for awhile. How many of those 10 key phrases do you hear? When you hear someone saying they're entering the traffic pattern, see if you can locate the aircraft visually. Watch them move around the traffic pattern as the tower gives instructions. It won't happen the first time you turn it on, or maybe not even the second. By the fourth or fifth time you'll probably hear someone using phrases you can understand. I'm sure you'll also hear a few of those folks that "think" on the mike, trying to figure out what to say next. If you live close enough to the airport, take your radio home and turn it on for 30 minutes a day. I guarantee within a few weeks, you'll find the entire process of learning the radio becoming easier.

There is one more phrase you can use when you begin using the radio yourself to make life a great deal easier: "I'm a student pilot, " before you begin your first transmission. This tells the tower you're new. They'll adjust their speech rate for you. Now I know many of you may think calling attention to your status as a pilot in training may be the last thing you'd like to do, but take my word on this one. This will make your relationship with ATC flow much more smoothly. As you gain in experience, you'll realize just how important that is too.

Before we move on to the phrases you'll hear most often, let's talk about the places that will require radio contact. At tower-controlled fields, you'll find one frequency for the ground controller who is responsible for the safe transit of traffic to and from the runways; another frequency will be dedicated to the tower controller who issues the landing and takeoff clearances and makes certain the traffic pattern in the air is flowing smoothly.

A larger airport might have a radar installation that allows them to control the aircraft in the air for a 40-mile radius around the field. This approach control and departure control is located in the same building as the regular control tower. Here men and women sit in front of large radar screens that allow them to safely separate aircraft from one another. Approach control and departure control will have separate frequencies, many times two or three frequencies apiece.

For weather information en route, you might contact the Flight Service Station (FSS), which is often located at an airport where there is no control tower. The FSS uses remote transmitters and receivers at other airports to communicate with aircraft, too.

Finally are the local airport advisory services provided on unicom, also located at airports without control towers. The unicom could be located in the FBO's front office with a set of wind instruments. When you call in, someone looks at the wind speed and direction and tells you what they see. They are not, however, air traffic controllers and they do not provide any type of separation between aircraft landing and taking off.

If you've been reading closely, you'll notice all ATC phrases, even though they relate to different kinds of facilities, have a distinct similarity. It wasn't an accident. No matter who you are calling on an aircraft radio, you can use the same format and get the results you want. The format is to tell the ground facility who you are, where you are, and what you are requesting.

Ten primary air traffic control phrases

Calling ground control.

1. Louisville ground, this is Citabria 9 Mike Kilo at the general aviation ramp for taxi to takeoff westbound.
2. Memphis ground, Bonanza Two Three Four Five just landed on Runway Eight, request taxi to the general aviation ramp.

Calling the tower.

3. Colorado Springs tower, this is Duchess Two Three Four Five, at Runway One-eight, ready for takeoff.
4. Danville tower, this is twin Cessna Two Three Four Five, ten miles east, landing with information Alpha.
5. Republic tower, this is Cessna Two Three Four Five, student pilot, eight miles east for landing with information Bravo.

(When the Automatic Terminal Information Service (ATIS) is broadcast at an airport, the pilot's initial communication with the controlling facility should identify current airport information. When there is a change in the weather or a change on the airport that might affect safe operations, the controllers update the ATIS broadcast.)

Calling approach control.

6. Springfield approach, Grumman Tiger Two Three Four Five is 25 miles west, overflying Springfield Airport at five thousand five hundred feet, requesting traffic advisories.
7. Gulf Port approach, Cessna Five Four Three Two is thirty miles east descending from four thousand five hundred feet, landing with information Delta.
8. Milwaukee approach, Piper Niner Niner One One, student pilot, twenty-five miles west with information Lima, unable to locate airport, requesting vectors.

Weather information.

9. Flight watch, Beech Five Five Five Five, student pilot, just south of Alligator Alley, VFR, en route to Fort Meyers, requesting Fort Meyers' weather.

Landing at an airport without a control tower.

10. Campbell unicom, Cessna Niner One Niner Niner, student pilot, five miles west, landing, requesting active runway and wind conditions.

It always works. Keep in mind though that learning radio lingo is like exercising. If you don't work at it often, you'll start to get flabby.

THE BLAHS

You now have an idea of what a typical lesson will be and you've learned how to avoid some of the pitfalls of talking on the radio. Each lesson you take is going to be different, with its own highs and lows. The main thing to remember is not to get discouraged when your progress seems to be slim or if you seem to regress. Believe it or not, this too, is normal.

There is a graph, called a learning curve (Fig. 3-2) that some incredibly smart psychologist put together to show the way in which the human mind learns new tasks. It makes no difference whether you're learning to program a computer, fly an airplane, or bake soufflés. The first few lessons you'll learn at a tremendous pace. Each lesson will prove interesting. After about six or eight lessons, perhaps about the time you are trying to perfect your landings when flying, you may find yourself coming out for a lesson and walking away feeling like you should give up the entire project.

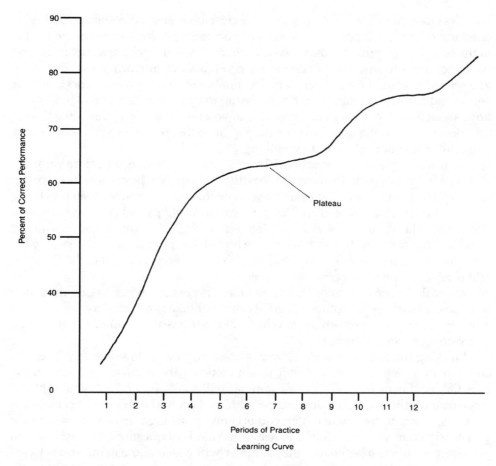

Fig. 3-2. A learning curve shows that slumps in training are normal.

But stick with it and eventually you'll pull out of the mental dive and begin learning new material again until the next flat spot. These flat spots are called plateaus. They're going to happen, so be ready for them and don't let them throw you.

Everyone has bad days and I'm sure as a student pilot you'll run into your own. Don't give up. Sit down with your flight instructor and explain your frustration. Instructors see the same problems over and over, and a good instructor is going to see the problem before you're even aware of it. You will feel bad when you make mistakes, so let your instructor soothe your ego to help you get back in there and keep flying.

PICKING UP YOUR LICENSE:
The flight test

Once you've made it through all the various phases of instruction for your private pilot license, including the written exam, you can begin thinking about the flight test. One day, out of what seems like the clear blue sky, your instructor begins checking over your logbook to be certain you meet all the requirements for the flight test. You've probably noticed, too, that as you fly your instructor doesn't have too much to say, except something good, about your progress. "Better get you set up for a flight test!" There it is. The final exam has been announced. All the weeks and months of work have finally begun to come together. If you have questions about flying that have been gnawing at you for a long time, this is the time to ask. Better to learn the answers now than to have the examiner make you realize you don't know or understand everything you should.

Remember final weeks at school? They always put them right before vacation. You couldn't move on to the next grade without them. No differences here, except on a flight test you'll be doing exactly the same maneuvers you've been practicing with your instructor. There is an FAA publication you'll probably want to pick up about now, the *Private Pilot Practical Test Standards* (PTS). This book lists all the standards required for the exam. There are no hidden, pop questions. If you fly the maneuvers satisfactorily and are familiar with the subjects covered in the PTS, you will receive your license at the end of the exam.

Every flight test examiner knows a student is nervous when taking the exam; some, more nervous than others. A good examiner takes your anxiety into account when watching your performance. What will the test really be like? What are you really going to see on the big day?

Working toward your license, you will already have flown several hours of cross-country time as well as spent time practicing the maneuvers listed in the PTS. On the day of your test, your examiner will meet you and ask you to plan a cross-country flight to some destination within a hundred miles of your home airport. The examiner will watch how you organize your flight logs and listen while you obtain your weather briefing to hear what kinds of questions you ask. As you prepare your navigation charts, the examiner will watch and ask questions to see if you really understand what you're doing.

When your flight log is completed and you're ready to head out to the airplane, the examiner will ask you to sit down and talk about flying. The questions now might relate to how much fuel the airplane carries or how much baggage. He'll ask what makes an airplane fly? What is the control tower frequency? How many takeoffs and landings must you make and how often to remain current? What are the weather minimums to fly VFR?

It might sound like this is a tough test, but this is material you should have covered with your instructor. This section of the test is called the oral and you must successfully complete it to move on to the flying portion. Remember though, if you are asked something on the oral that you honestly do not know, say you don't know. It is not uncommon for an examiner to press harder in an area a student seems confused about. That pressure can make a student talk about things they really do not understand. This could mean the end of the flight test. A wrong answer, however, does not automatically mean you've failed because it depends on what the question is. If you give the examiner an incorrect radio frequency, it is obviously not as bad as if you told him you don't know how much fuel the aircraft holds or how much fuel it burns per hour.

When the oral is complete, you'll know because the examiner says, "Let's go out and preflight the airplane." Now the examiner will watch your preflight to see how careful you are. Expect the examiner to ask more questions about what you are doing. Which way does the elevator move when you pull back on the wheel? How many cylinders in the engine? How do you know if your aircraft has been fueled correctly?

When you finally enter the airplane, the examiner will expect you to perform all the functions yourself, with absolutely no help. As you start up and taxi out, the examiner's criteria are simple. There must never be a time when you are not in complete control of the airplane. If that happens, the exam is over. As you take off and climb away from the airport, the examiner will continue to ask questions.

By now you might be realizing what a distraction this passenger is becoming. This is no accident. In the new tests given to pilots, the examiner is expected to distract you to watch how your control of the airplane changes as unexpected events occur.

No one expects you, as a new private pilot, to be able to easily coax a large crippled airliner safely to the ground, but they do expect you to be able to cope with a radio failure while you're approaching the airport for landing. (By the way, I still wait for someone to announce their need for a pilot on the P.A. every time I fly on the airlines. Hasn't happened yet though!)

The flight test will begin with the cross-country flight you planned earlier. Once the examiner realizes you are on course and have successfully located the first few checkpoints, he'll tell you to cancel the flight plan and head away from your original destination to perform some airwork. This is when the examiner will ask you to demonstrate slow flight, turns, stalls, and ground reference maneuvers.

Don't be concerned if the examiner says: "Let me show you one of these." Just release the controls to the examiner and watch and listen. The examiner most likely will demonstrate one and offer you a chance to try again. When the airwork is complete the examiner will tell you to head back to the airport for some landings. You'll be expected to find the field, use the radio, enter the pattern and make whatever landing the examiner requests, be it short field, soft field, etc. Don't be surprised if, during one approach, the examiner suddenly says, "A truck just pulled on to the runway. What do you do?" The examiner expects you to perform the go-around maneuver and make the aircraft begin climbing.

At some point, the examiner will say, "Let's make the next landing full stop." Don't get nervous but this is most likely the last landing you'll make as a student pilot. Most examiners don't like to talk until you have returned to the ramp and shut down the engine. If you've made it this far OK, they should reach over and congratulate the newest private pilot. It is permissible, after you leave the airplane, to jump up and down and yell your head off.

You're a pilot (Fig. 3-3).

I. UNITED STATES OF AMERICA   XI.

Department of Transportation – FEDERAL AVIATION ADMINISTRATION

THIS CERTIFIES   IV. ALAN JAMES FOSCO
THAT   V. 1433 E. KENILWORTH
                       PALATINE, IL   60067

DATE OF BIRTH	HEIGHT IN.	WEIGHT	HAIR	EYES	SEX	NATIONALITY	VI.
07-22-66	66	145	BROWN	BROWN	M	USA	

IX. HAS BEEN FOUND TO BE PROPERLY QUALIFIED TO EXERCISE THE PRIVILEGES OF
II. PRIVATE PILOT   III. CERT.NO. 325828194
                    RATINGS AND LIMITATIONS
XII. AIRPLANE SINGLE ENGINE LAND

XIII.

VII. _____Alan J. Fosco_____
            SIGNATURE OF HOLDER   X.
X. DATE OF ISSUE: 06-12-89   VIII.   *T. Alan McArtor*
        AC Form 8060-1 (8-87) Supercedes previous edition   ADMINISTRATOR

Fig. 3-3. The prize at the end is a private pilot certificate.

4

Now that you have
a license . . .

YOU'VE JOINED THE RANKS of the licensed pilots of the world. In the United States alone there are more than 692,000 active pilots, flying everything from the giant Boeing 747 for major air carriers down to a tiny two-place hand-made experimental aircraft. Where do you fall amongst these groups? Now that you have your license, what are you going to fly?

The easiest solution is to continue renting the same airplane you've been using for your training. The owner or the FBO will usually be glad to have you stay as a renter because you're a known person. They trained you and know exactly what your capabilities and limitations are. In fact, I recommend you do just that for the next 30 or 40 hours—don't change airplanes.

You've been through a lot of work and the first hurdle, your license, has been passed successfully. It's time to give yourself a break and enjoy, really enjoy, some flying time. It might mean taking an afternoon off from the office and heading out to the airport to fly away for some solo time (now called pilot-in-command time because you're rated in that airplane) to see the sights. Perhaps you'll head over to that airport 30 miles down the road and work on your landing technique. Either way, it's your time to explore, to let your mind run free and consider all the things you'd like to do in the airplane, the places you'd like to see. Just loosen up and fly. Take your spouse up for the first ride. Or take your kids. Show them the house from the air. Show the town and their school. Explain what you're doing as things happen. You're a pilot now.

I made a regular booking of the Cessna 150 I picked up my license in for every Tuesday and Friday morning before I would head off to work. To me, being up an extra hour and a half early was worth it. I'd arrive at the airport just after the sun came up. The place was still quiet except for the occasional whine of the turbines from corporate jets on the other side of the field as they started up for another day's work. The air was still and smooth on departure and looking eastward I could see the sun brilliantly reflecting off Lake Michigan and the downtown area of Chicago.

Some mornings I'd head northbound toward Milwaukee, sometimes eastbound to the lake, or south toward Indiana. Other mornings, depending upon my whims, I'd fly west toward acres and acres of the golden corn fields that cover Illinois. I always wished I lived in other towns so I could fly my morning patrol around the New Orleans area of the Gulf of Mexico. It would be so nice to watch the Rampart Range of the Rockies out the window as I took off from Colorado Springs. Then of course there was the tour of the Pacific Coast I wanted to take to duplicate the run down Highway One I made in my car a few years back, from San Francisco south. Maybe you live in some of these places and never realized the value of the scenery around your home. I think it's time you did. Go on, get in the airplane and explore.

It takes a good 10 to 15 hours of flying before you begin to overcome the fear that you might have passed the checkride by accident. You know you took the test and you know you passed it, but sometimes that message really doesn't get through to every part of your brain. Sometimes, you still can't believe that you're really a pilot. Now the initial shock has worn off and you realize that you really are the only person in the airplane. So yes, you must be the one flying! You'll find yourself still going through various stages as you did when you were just a student. Some will be good plateaus and others will involve times when you'll end up possibly scaring the heck out of yourself. Just so you don't think it happens only to new pilots, here's one of mine.

PLATEAU OF ARROGANCE[1]

When I was only a mere pup of a student pilot, an instructor dutifully warned me of the various plateaus in flight training. There will be a time, he said, when you'll have six or seven uneventful solo hours under your belt, and you're going to do something stupid and scare the hell out of yourself.

Sure enough, I did.

He also told me to watch for the couple-of-hundred-hours syndrome. That's when you saunter around flight operations and reply with an "I've almost forgotten how many hours I really have" nonchalance to those who seem to covet logged hours. Who would have thought I would be looking back at another plateau in the nonchalance cycle of piloting, as I am right now, thinking how lucky I am to be alive to tell the story.

I was managing an FBO flight department. The job and I were made for each other, and I arrived in my new office with most of the flight instructor ratings as well as an ATP and 4000-plus hours. Because of the extremely heavy charter load we were having that summer, few people were around the airport to test aircraft for the maintenance shop.

---

[1] "Plateau of Arrogance," reprinted from *Flying*, August 1983, by permission.

I fell into that duty with enthusiasm. It gave me a chance to fly a little by myself and got me away from the office for half an hour every so often.

Our shop foreman strolled into my office one humid July afternoon to inquire whether I could take an old Aztec out for a test flight to check the operation of the right engine. I peered through the blinds of my office and spotted an ancient Piper on the ramp. It was one of those very old Aztecs that looks like an Apache with a big tail and longer nose. I knew full well I'd been in an Aztec only twice in my life, and one of those flights had been in the back seat. But wasn't I a multiengine pilot? Wasn't I a multiengine instructor? Wasn't I extra current in the Cessna 310, the Cessna 400 Series and Barons? Besides, I was an ATP.

The Aztec's owner, a man in his early 70s, wanted to come along, but was quite happy to have me fly the machine. "Imagine, the flight department manager taking the time to check out my airplane," he said. "What service!"

I had some trouble getting the engines started—until I managed to find the master switch. I waved to the shop foreman and called DuPage ground for taxi. As I taxied to the runway, I noticed the airspeed indicator. It was so old, the damn thing wasn't even marked with the V_{yse} speed (best speed to climb at if one engine quits). I was cocky, but I wasn't anybody's fool. I queried the owner. "Oh, blueline's about over there," he said, pointing to a spot near 100 mph.

It was about this time that I began to feel those first small, almost imperceptible beads of sweat break out on the back of my neck, telling me I should call a halt right there and get the facts. But then, wasn't I an ATP and a multiengine flight instructor? I proceeded to the engine runup. There was no checklist.

I knew the fuel pumps had to be on for takeoff and landing, and the old CIGAR checklist (Controls, Instruments, Gas, Altimeter, Radios) seemed to work out okay. When I got to the autopilot, I saw that the switch to turn it on and off was a push-pull type attached to a dull silver ball bearing knob. There were no markings anywhere around the thing, so I queried the owner again. He told me that "push" was off. I nodded. I had no need to worry about the heading bug because the autopilot was off, so I just left it where it was—due west.

I brought in takeoff power as we rolled down Runway 15. No sooner had we lifted off than the aircraft started a steep turn to the right. It took both hands to hold it even close to level. My eyes, not accustomed to the instrument placement, darted from place to place until they landed upon the manifold pressure gauge for the right engine. It was down to about 16 inches. Jeez! We're losing the right engine, I said, loud enough to be heard over the noise of the wind and the good engine. We now had about 100 feet between us and the ground, and the tower was yelling to know if we needed help.

I told my copilot to turn down the radios. I was on my own now. Just as I was about to reach up and feather the right engine, I noticed that the right throttle was retarded halfway back on the quadrant. Obviously the friction lock hadn't held when I let go of the throttles to grab the control wheel. I immediately pushed it forward to find the manifold pressure return to normal. I had two good engines again, but the aircraft was still fighting me to head west by itself. WEST! My right hand dove for the bug on the autopilot and turned it to the 190 heading we were passing through. The aircraft immediately righted itself. I didn't touch another thing again until we had 1000 feet under us.

At 2000 AGL (Above Ground Level) I turned the bug on the autopilot and the aircraft responded. Halfway through the turn, I pulled the little silver knob out from the autopilot and the system disengaged. I glared at the owner. He shrugged his shoulders and looked at me sheepishly.

By this time I was dripping with sweat. I turned the aircraft around and landed. As we taxied in, I wasn't sure whether I should kick myself, buy a hairshirt or submit to 20 lashes from our chief pilot.

With all the experiences I had behind me, the thousands of hours, when it came right down to it, all I'd reached was another plateau of arrogance about my abilities. As I drove home that night, I vowed to try and remember that I was, even with 4000 hours, just a pup. A little bigger than some, but . . . just a pup.

You will indeed make mistakes as you learn. The most important piece of advice I can give you is to learn from them.

WHAT TO FLY

Because you've joined the ranks of licensed pilots and have gained another 30 or 40 hours since your flight test, making local and cross-country trips, you're probably ready to consider flying another type of single-engine airplane. But which one?

The easiest way to determine which one you would like to fly is to decide what you need from a different airplane. Perhaps you have a family of four and a two-seat Cessna 150 just won't do the trick, or you think it's too slow. The best suggestion is to sit down with pencil and paper and start calling around to various places that rent airplanes. Find out what other aircraft are available and at what cost. Very often you'll find an ad in the newspaper for someone who is willing to rent their airplane.

Many factors should be considered when renting an airplane. Let's cover a few. The cost of the airplane is usually the first item on anyone's list. Why will two Cessna 172s vary by $5 or $10 or even $15 per hour to rent? First of all, let's use a basic rental cost of $60 per hour, which is not too uncommon in 1993, for a four-place Cessna 172.

You see an ad for a Cessna 172 that says you can rent this airplane for $50 per hour. Your first thought is that this is one incredible deal so you'd better jump at it. As you call, you find out from the person that this is a dry rate, meaning you must pay for the fuel. Most people rent aircraft at a wet rate, meaning the cost of the fuel is included in the hourly charge. If you make a cross-country trip in the plane, and refuel along the way, you bring the receipts back and that cost is deducted from your total bill for renting the airplane. People will advertise a great rate like $50 and often not mention that it is a dry rate. You'd have to know enough to ask.

A Cessna 172 burns about eight to nine gallons of fuel per hour and depending upon where the fuel is purchased, the cost per hour could vary from $12 to $17 per hour in addition to the $50 per hour rental rate. This "cheaper" 172 has become a more expensive airplane than the first one, by as much as $7 per hour.

All Cessna 172s are not equipped the same (Fig. 4-1). One Cessna 172 might cost $47 per hour but be relatively bare. A magnificently equipped 172 might run $60. It means you, the renter pilot, must be on your toes to decide what an hourly rate really entails.

Fig. 4-1. A four-place Cessna 172.

For my money, I'll take the more elaborately equipped airplane every time. The reason is that a new pilot should expose himself to working with various types of new equipment as much as possible to gain as much experience as possible. These days there is an updated version of some piece of avionics entering the airplane market almost monthly. Every piece is designed to make flying easier as well as to help you squeeze every extra ounce of performance from an airplane. Take advantage of them when you have the chance.

The age of an airplane is another factor in the cost of the rental. When someone rents an airplane, they tell you a Cessna 172 rents for so much an hour and they might tell you how it's equipped, but they seldom tell you the age. Now age isn't usually as important in flying as it is in an automobile, because airplanes are maintained to much higher standards than automobiles. A 1975 172 will probably be as good as a 1985. However, I believe there is a limit to this kind of thinking. The owner of a 1957 Cessna 172 can put an ad in the paper and say "Rent a Cessna 172 for $29 per hour," and you'd think it was a heck of a deal.

Even with all the time I have flying other people's airplanes, I was recently so involved with a telephone conversation that I forgot to ask the age of the airplane.

This man told me his Cessna 210 (retractable gear, about 190 mph) contained every labor-saving device known to pilots short of a maid and a bar. It sounded too good to be true for the price that I should have known it was just that. When I arrived at the airport I saw what looked to be the oldest, dirtiest Cessna 210 ever put on God's earth. The owner tried to downplay the cosmetics. I'm a firm believer though that if the owner doesn't take care of the outside of an expensive airplane, the chances are good the same applies to the important parts inside too, like the engine and landing gear. I said thanks, but no thanks.

Remember, when you rent an airplane from an individual, as opposed to your original flight instructor or FBO, you really have no way of knowing how the airplane has been maintained. Before I fly an unfamiliar airplane, I always ask who the regular mechanic is and then spend a quarter for a phone call to learn something about the airplane and its owner. If the owner should balk at this idea, consider another airplane. This could save you a great deal of grief in the long run.

Consider the insurance that is carried on the aircraft you fly. Make certain you see a copy of the policy. Most carry two deductibles, one while the aircraft is stationary and the other while it is in motion. As a renter pilot, should damage occur to the aircraft while it is your responsibility, you are liable for an amount at least equal to the deductible. In some light singles, it could be $100 not in motion and $500 in motion while a light twin might reach $1000 in motion deductible.

So what's the answer?—renter's insurance. Several aviation insurance companies offer renter's insurance for pilots. The basic policy might cover the deductibles on any airplane you fly and also provide additional liability insurance, which is always a good idea.

Now that you have an idea of what to look for when renting, which airplane do you want to fly? There are many to choose from almost anywhere you live. Depending upon the type of aircraft available, one or two of the next bigger models in that manufacturer's product line will satisfy the customer who wants to upgrade to bigger or faster. Examples would be an operator who rents a Cessna 152, 172, 182, and 210.

Possibilities include moving up from the two-place trainer to a four-place single engine with slightly more horsepower or perhaps extra equipment installed. Another option is to move up to an aircraft that has retractable landing gear and a variable-pitch propeller for extra speed. All of these will carry extra baggage, as well. There are six-place single-engine aircraft that feature either fixed or retractable gear. You might want to move from a high-wing airplane to a low-wing or vice versa. The combinations are almost endless.

No matter what you do decide, you'll find a checkout flight is necessary before flying the aircraft by yourself, especially if you are a low-time pilot. Less than 100 hours total time is considered very low time, but don't be disheartened. It will build.

An upgrade checkout can be a simple matter if you were flying a Cessna 152 and move up to a Cessna 172 because the aircraft are very similar in design. The 172, in fact, looks very much like just a slightly larger 152. The 172 flies faster with a bigger engine and carries more payload. And that's really the same idea that carries forth if you begin in a small Piper or Beechcraft. The upgrade models incorporate certain similarities to make it easier for you to move up.

The checkout will be designed to show the instructor or owner that you really understand the difference between what you were flying and what you plan to fly now. This involves some ground school to show you the differences as well as a written test to prove the new airspeeds and weights have been committed to memory. After the ground school, you'll go out and fly to give your instructor the opportunity to introduce you to both the differences and similarities in flying and landing characteristics.

When the instructor is satisfied you're capable of handling the aircraft by yourself, you'll be on your own again, another step up the ladder. The checkout could take longer if you are a low-time pilot trying to move directly up to a retractable-gear airplane. This type is considered quite complex by most insurance companies, which might require at least 10 hours in type with an instructor along before solo, which can become expensive. It usually makes more sense to move up slowly from type to type, which shows the insurance company that you're not biting off more than you can chew.

INSTEAD OF RENTING

If you've ever rented a car, you know the process isn't always as easy as the rental companies would lead you to believe. Often you'll arrive to find the reservation has been lost, no car is waiting, or the model you wanted is out and only a smaller model is available. Just like auto rentals, aircraft rentals can have specific problems.

Airplanes are machines and sometimes break. You might arrive at the airport with baggage and kids in the car ready to leave for an exotic weekend getaway. You might find the aircraft you reserved has a flat tire, broken radio, or an oil leak that makes it unairworthy. Unlike an auto rental firm that will give you a larger vehicle if the one you asked for is out of stock, the aircraft rental business—with its higher initial investments and tighter profit margins—most likely won't do the same. What is often more frustrating is to find that an aircraft has been out of service for two or three days and no one bothered to call you. Call the rental desk or check the calendar yourself 24 to 36 hours prior to the reservation to verify availability.

What's the solution? Consider buying an aircraft of your own (Fig. 4-2). Too expensive you say? Maybe, but then again, maybe not. Certainly, owning anything today is going to cost more than it did in 1979 but less than it will 10 years from now. Cost is relative.

So how do you decide if you might be a candidate for ownership? Quite simply, research. Before you buy, or decide to buy, an aircraft, pick up all the facts you can and then look at your budget to make the decision. Don't let emotions run away with you. You might end up buying when you should continue renting, or possibly end up buying an aircraft that is too expensive for your budget.

Aircraft are, in a sense, very much like automobiles. Some owners take meticulous care, while others just barely perform the work that needs to be done. The trick to figuring out aircraft is talking to the experts: owners and mechanics.

New terminology can be illustrated in a sample ad that you might see in publications that offer aircraft for sale: "1976 Cessna 172, 800TTAE, Dual ARC Nav/Com, G/S, A/P, 3LMB, EGT, Xpndr, Encoder, ADF, NDH, hangared."

Fig. 4-2. A four-place Piper Cherokee.

The first portion of the ad is straightforward with the year, make, and model.

800TTAE means there is 800 hours total time on the airframe and the engine. This also tells you the aircraft has never had an engine overhaul. An aircraft that was used more—sometimes referred to as a high time Cessna—might show 2800TT, 900SMOH. This means the aircraft has flown 2800 hours since its date of manufacture and has flown 900 since major overhaul of the engine.

Aircraft engines are given a recommended number of hours to fly before the manufacturer believes they should be torn down, checked, and new parts added to keep them running smoothly. The average length of time is between 1800 and 2000 flying hours. The advertised example aircraft will be due for another engine in about 900 hours. The *TBO* (time between overhauls) is an engine manufacturer's recommended time and not a legally binding figure. You might find aircraft flying that are 200 hours past overhaul. While the engine might seem sound and not burn oil, a prospective purchaser would have to ask how long that could go on. Ideally, the engine should be considered run out.

Dual ARC Nav/Com refers to the basic radio package included in the aircraft. There are two radios: one has a G/S (glideslope) receiver, which will become more important as you move on to the instrument rating. ARC stands for Aircraft Radio Corporation, a company formerly owned by Cessna Aircraft, and quite a common radio among this breed. Other radios in a classified ad might be King, Narco, or Collins.

A/P means the aircraft has an autopilot. The brand name is not shown. It could be a simple autopilot that operates around one axis and keeps the wings level. A more sophisticated unit would include controls built into the aircraft to also hold altitude and heading steady. These units would increase the aircraft's total value.

3LMB means the aircraft is equipped with a three-light marker beacon, another piece of equipment that comes into play during instrument flying.

EGT stands for exhaust gas temperature gauge, which helps the pilot lean the fuel and air mixture in the engine for the best possible performance.

Xpndr is the transponder, which helps ATC identify the aircraft on radar.

Encoder stands for the altitude encoder. This is an electrical device attached to the back of the aircraft's altimeter and then into the transponder to give ground based air traffic control radar your aircraft's altitude on a regular basis.

ADF is the low frequency radio used for navigation. It operates in the band beginning about 200 kHz to 1600 kHz or the end of the broadcast radio band. Yes, this means you can listen to the LA Rams game in the air if you know the right frequency.

NDH is a mighty important one since it stands for No Damage History. This is not required in an ad, but it's one of the first things I'd want to know about when I called on an airplane. A mechanic is the person to be able to tell you if any damage listed would be significant or not.

Hangared means the owner kept the aircraft in a hangar. If this is the original owner, you know it has been covered since it was new. This means the paint is most likely in good condition.

In a recent copy of *Trade-A-Plane* (410 W. Fourth St., Crossville, Tennessee 38557) there were 295 Cessna 172s for sale. At first glance you might think there is something wrong with this model aircraft because so many were for sale at one time, but that assumption is only to the untrained eye. In actuality, more Cessna 172s were produced than just about any other single-engine airplane in the world. If you happened to be searching for a Bellanca Citabria—a fabric covered aerobatic aircraft—only 20 were advertised this same week. It just depends upon what you're seeking.

Ads run with phone numbers from every state and Canada. If you live in New York, you should narrow the search to those aircraft within a few hundred miles of home or costs to view the aircraft will climb out of sight. If, however, you were looking for one of those rare aircraft, you might expect to spend more for search costs because the best ones might be in Southern California.

Because you're a pilot, you'll find the time and money you spend to rent an airplane and fly out to see an airplane will be money well spent. An aircraft located two hundred miles away can be viewed in one morning instead of a full day. Before you look for an airplane, find a competent mechanic and talk about your plans. When it comes to mechanics, every pilot thinks his is the best, so selecting one can be tough. The best way is through a referral from someone you know who owns an airplane. If you don't know anyone directly, perhaps the local pilots' association can give you a few names to call for information. If none of these methods work, get out the trusty legal pad again and start calling the local FBOs, much the way you located a flight school, and offer to trade the mechanic lunch for some advice.

One mechanic might tell you to watch out for certain 172s because they suffer from chronic engine problems. Another might tell you of brake or gear problems in certain models. No matter what you hear, write it down, as well as your opinion of how helpful that particular mechanic was. If all they told you was to call them after you bought the airplane, scratch them from your list of prospective shops. If you feel good about what the mechanic says and you feel he really cares about whether or not you come back, you might have found the right shop.

Before you buy any airplane, you'll want to have a pre-purchase inspection performed. This is like taking a used car to a mechanic for the straight information on what is wrong and right with the machine you're considering. Depending on the airplane, you might spend a few hundred dollars on the inspection, but again, it's money well spent. It's certainly cheaper to learn the airplane in question has a bad cylinder or a frayed aileron cable before you buy it, so you can negotiate the price (Fig. 4-3).

Fig. 4-3. A Beechcraft four-place aircraft, the Sundowner.

Aircraft operating costs must be considered: direct, fixed, variable, and reserve costs. Direct operating costs usually include the fuel and oil used per hour. Fixed expenses cover the items such as insurance, mortgage, and tie-down fees at the airport. Usually fixed expenses do not change, but these days many banks are financing aircraft with the same variable interest rate loans so popular among homeowners. While these rates are usually tied to 91-day treasury bills and can change each 90 days, a word of caution is necessary.

Before you accept an initially lower interest rate on a loan, be certain you realize just what the cap or the top end on that loan might be. Always calculate your operating costs on a worst case scenario. You might find that should the interest rate climb to 15 percent, your dreams and your airplane could fly right out the window.

Another word of caution is necessary about banks that offer variable interest rate loans but tell you that the payment will never change. In this case, you might end up owing more on your aircraft loan than you originally borrowed. If the payment remains the same but the interest rate goes up, the bank will take less and less of your payment and apply it to the principal on the loan.

There is indeed a point where the interest rate can be high enough that there is not enough left of your payment to pay the interest on the loan each month. In this case the bank will add whatever amount is necessary to the balance of the loan to cover the interest charges. In the end you could end up owing hundreds or even thousands more on the loan than anticipated. Negative amortization loans are dangerous and should be avoided.

Next comes the variable operating expenses: money you'll need to spend over the course of the year to repair things like a blown tire or a broken radio. While these are tough to plan for, you can expect to pay more money each year if you have more options on the aircraft because the opportunity for something to fail is greater. This will also depend upon how the aircraft was maintained by its last owner. You might pay a slightly higher price for an aircraft that was maintained better, but the extra few dollars you add into your fixed cost mortgage payment each month could be dollars well spent. The systems will probably not break down nearly as often.

Finally, cash reserves that should be set aside for inspections and overhauls. Aircraft, unlike automobiles, must meet a fairly simple, but sometimes expensive, inspection schedule. Each 12 months the aircraft must be fully inspected to determine its fitness for flight. This inspection can be performed by any number of maintenance shops around the country. You can shop for the best price and results by talking with other aircraft owners around your airport to determine costs and the general reliability of the shop you choose.

How much will the inspection cost? That's tough to say exactly. If you had a prepurchase inspection performed before you bought the aircraft, there should not be any major surprises like a dead cylinder or corrosion in the fuselage. Inspections are sometimes sold on a flat fee per aircraft basis plus a per hour cost if anything unusual pops up. Around the Chicago area, an annual inspection on a clean (no major problems) Cessna 172 might run about $700 to $800, while in Colorado Springs the same inspection might run only $600.

The most important consideration is to add a few dollars into the "kitty" for each hour you fly the plane. The "kitty" will cover the costs of these inspections and overhauls. In addition to the regular annual inspection of the aircraft required by FAA, an aircraft engine must be overhauled. This is a major undertaking. It involves completely pulling the engine off the aircraft and tearing the engine down to determine what if anything needs to be replaced, such as crankshafts, pistons, magnetos, etc.

On an aircraft like a Cessna 172 this could cost $8000 to $10,000. However, depending upon the model 172 you buy, the overhaul might not come until 2000 flying hours have been accumulated. If you buy an aircraft with 800 total hours on the aircraft and engine and use the aircraft only 200 hours per year, you could very well be years away from the overhaul. This will allow plenty of time to begin the reserve fund to cover the project. If an overhaul costs $9000 it is simple to determine that since new, you should have been putting away $4.50 per hour to cover that engine work ($9000 ÷ 2000 hours = $4.50).

If you purchased the aircraft at 800 hours, you have to make an adjustment in the hourly costs to compensate for the time you didn't own the aircraft and make it equal $9000. This would add another $3 per hour to the $4.50 you already planned. Your engine reserve is now $7.50 per hour. Keep in mind that no one will be standing behind you, forcing you to drop $7.50 into the bank each hour you fly your aircraft, nor will they tell you to allow another $1 for avionics repair or another $3 per hour to cover the annual inspection. In general, radios don't need much attention unless something is wrong, so your reserve for repair can be much less.

You could fly the airplane and pay as you go. In this world of payment schedules, most people find it easier to keep track of where the money is going and to plan for upcoming expenses by breaking everything into an hourly figure.

Let's take a look at actual costs to show the difference between owning and renting the same airplane. If you've been renting a four-place, single-engine aircraft such as a Cessna 172 at $55 per hour and fly it 100 hours per year, your cost is $5500. Add another $100 for renter insurance and the grand total for flying is $5600.

If you had purchased a Cessna 172, you probably could have bought a fairly nice one for $18,000 to $20,000. Based on 12.75 percent interest and a seven-year repayment schedule, that aircraft would cost $263 per month with a 20 percent down payment. Insurance comes to another $600–$800 per year, depending upon where you live and how the aircraft is equipped. You need a place to keep the aircraft, so a tie down will range from $35 to $85; you can easily save $25 per month on a tie down at a remote field. You probably wouldn't use the plane as much if you had to drive an hour to get to it, not to mention the auto gas expense.

Payment	$263
Insurance	$ 60
Tie down	$ 50
TOTAL	$373

These monthly fixed costs must be considered before buying the aircraft. Fuel will cost about $12 to $14 for each hour and oil about another $.50 per hour. To be certain you don't run short of cash, should something break, you can set aside an amount for each hour you fly as a maintenance reserve. This might account for another $7 per hour.

This is beginning to sound a bit more expensive than the rental, and in a way it is; in another way it isn't.

If you flew someone else's airplane for 100 hours per year, you probably found there were trips canceled due to weather. Or perhaps someone else didn't show up with the airplane when they were supposed to because of some problem at the other end. The supply of excuses you might hear when the airplane isn't where it should be can be endless.

Let's make an assumption that if you had your own machine you might fly more often. You might find yourself sneaking away from the office to catch an hour after work on a nice night. How about taking your wife and kids to dinner at a resort that would be an impossible car trip. Maybe you'll head out to the airport on a nice Sunday morning. Take your neighbor or your son and explore the far away regions of your state for a few hours or so. Assume you are going to be a bit more adventurous than you were when you rented from someone else and you're going to fly another 50 hours per year. Let's look at all the costs again.

- One year's aircraft rental at 100 hours = $5500 (150 hours = $8250).
- Cessna 172 purchased for $20,000 flying 150 hours per year.
- Cessna 172 fuel at 7.5 gallons per hour at $1.75 per gallon = $13.13 per hour.
- Oil at $2.50 per quart = $.50 per hour. (Oil can be purchased a case at a time for a significantly lower price instead of by the quart.)
- Direct costs of $13.63 per hour.
- Variable costs $4 per hour.
- Engine reserves of $7.50 per hour.
- Avionics reserve of $1.50 per hour.

Total of variable, direct, and reserves = $26.63 per hour.
If we average out the total fixed costs per year:

- Payments = $3156
- Insurance = $1080
- Tie down = $ 720

Total $4956 ($4956 ÷ 150 hours = $33 per hour)
Total hourly rate = $59.63 per hour ($26.63 + $33)

If you were to be very enthusiastic and fly 250 hours per year, the costs per hour drop even more. Direct costs remain the same at $26.60 per hour, but fixed costs drop: $4956 ÷ 250 = $19.80 per hour. It actually costs you much less per hour for each additional hour you fly. Total hourly rate (250 hours per year) = $46.40.

Total cost for a year at 250 hours is $11,600. If you flew the same 250 hours in a rental aircraft, you'd spend about $13,750, almost $2200 more. It's like picking up an extra 46 flying hours per year for free. Besides, you own the airplane!

So, for essentially what you are paying per hour for someone else's Cessna 172, you could own your own and never again face the uncertainties of whether or not the aircraft will be there when you want it. You won't have to rearrange your vacation to Colorado because someone else needs the airplane right in the middle of the week. You'll never again be wondering just how the machine was cared for. You'll know every step of the way because you were there to watch.

Certainly, too, there is the significant factor of aircraft appreciation. In the years since 1982, annual production of general aviation aircraft has dropped significantly to the point that in 1988, slightly more than 1000 new aircraft were built. By 1994, the number will have dwindled to a few hundred. Most of these aircraft were turboprop and jet aircraft for the corporate and regional airline market. Only a fraction were single and twin engine piston aircraft for the private user.

In fact, in December 1988, Cessna Aircraft Company, once the dominant force in general aviation production, confirmed that it would not be producing any new single-engine piston-powered aircraft until many economic conditions changed. Beech Aircraft Corporation still builds a few aircraft each year while Piper Aircraft Company moved through Chapter 11 bankruptcy. Mooney and Commander still produce four-place single-engine aircraft however. But they all have a long way to go to catch up with 1979's aircraft production, when more than 17,000 general aviation aircraft were delivered.

LEASEBACK

Now that you've become embroiled in the facts and figures of owning an airplane, you might have decided it's too expensive for you. There is another aspect of ownership to consider: the leaseback.

Quite simply, in a leaseback situation you purchase the aircraft and lease it to an FBO who will then put it on the flight line as a rental aircraft. This kind of program has advantages for both the aircraft owner and the FBO. The FBO picks up the use of a machine he might not ordinarily have and can generate revenue for you, the owner, while you realize any tax benefits that owning a small business can produce.

The venture spreads the risk of operating an aircraft commercially between two parties instead of just one. While certainly the tax benefits of a leaseback are not as great as they once were, a leaseback can still be a viable source of revenue. It can help you offset some of the costs of ownership if you go into the project with some knowledge and your eyes wide open. Don't expect to buy and operate an aircraft you could not afford to own without the leaseback, however. If you should happen across an operator who tries to make you believe you can, run hard and long. You might have to pay for that aircraft someday, with no rental income, so make certain you can afford it.

Remember, a leaseback is essentially a small business. If you have no experience in business, you should consider some night school business education or a partner or an accountant who can provide the expertise to help keep you out of trouble when phrases like tax credits, lease payments, or depreciation are mentioned. Also consider why you should want to own a business. Your main reason should be to make money. If this isn't the main reason, and four out of the first five years you show a loss on your company's bottom line, you risk the IRS interceding upon the taxpayer's benefits and declaring your business to be a hobby.

If they decide you're involved in the leaseback business only to get a cheap airplane, you could lose most of the tax benefits you set out to gain in the first place. This could saddle you with a pretty hefty tax bill when you least expect it.

As a leaseback owner you must understand that a certain amount of trust must exist between you and the FBO. Today, trust is sometimes one of those esoteric phrases that contains a great deal more fluff than substance. A good business deal takes two people to run properly. If you don't know much about airplanes, find an expert. Pay him or her, if you must, to teach you about aircraft ownership from brake pads to major overhauls so you won't be left in the dark.

Finding an expert might not be as difficult as you'd think at first. Once you begin talking to the various FBOs around the field, you can begin asking them about the other people who have leased aircraft back to them over the years. Normally the operator should not have a problem providing the names of some of these people, but if he does, there must be a reason. The one usually vocalized is that the FBO does not give out that kind of information. A prospective leaseback owner should be highly suspect of an operator who refuses basic help, strictly on policy. I'd think the operator had something to hide, like an unhappy former leaseback owner somewhere. The best advice at that point is move on to another FBO with your leaseback plan.

These days, the most important aspects of the leaseback are the written lease itself and the insurance coverage as it pertains to that lease. Without these two items the rest simply doesn't work. The lease should spell out some very important things. For example, who is the owner and who is the operator. It should mention the time length of the lease and who can fly the aircraft. What should be charged per hour for use of the airplane, who pays taxes, and what are the various parties' obligations in various situations, such as if the airplane breaks down somewhere, are other considerations. How can you cancel the lease if you need to?

Certainly something I would want in my lease would be the right to inspect the paperwork that pertains to my aircraft. It's very simple for the shop to spend one hour changing a battery in your airplane, but you might find unnecessary repairs performed on the airplane. This is one reason it's so important to either be somewhat familiar with aircraft or know someone who is. If that battery change turned out to be a five-hour, $312 bill, you'd at least know enough to question it.

Something you might negotiate in the lease is a discount to the normal shop rate when the aircraft is worked on. Some shops will and some will not. The most important part of the lease, for maintenance, is to be certain you have some control of those maintenance items being repaired. You might decide that anything over $100 requires your approval before it can be completed and that only a few under $100 repairs may take place in a given month.

No operator (at least no honest one) is going to knowingly cheat his leaseback operator because he derives the benefit of having your aircraft to use. You cannot control the shop foreman, but $60 to change light bulbs or $600 to put new brake shoes on a Cessna 172 need to be caught and stopped, as soon as possible. There should be a provision in your lease to cover either party trying to pull a fast one on the other, although I'm sure the attorneys wouldn't word it quite that way.

In essence, most leases operate the same way. The FBO rents the aircraft out to the public for $50 per hour, for example. The lease might agree to return $40 per hour. If the airplane flies 30 hours in a month, you've earned $1200. Fuel and oil (if it was a wet lease) will be deducted from this payment, as will the tie down and any repairs needed to keep the aircraft airworthy. These various expenses, as well as your payments, depreciation, and insurance costs are all deductible on a Schedule C under ordinary business expenses. It sounds like a match made in heaven. And, it might be, until those winter months come along and the airplane's monthly income shrinks to some tiny sum like $125 because of the weather, while fixed costs remain the same. Like any other business, you need enough extra capital to see you through the lean times.

Have you thought about what you're going to feel when you want to rent your own airplane and the FBO says it's busy? Some of the stardust might fade from your eyes too when you find fresh grease ground into the carpet. Remember that the leaseback airplane is now in commercial operation and as such it will need to pass a regular 100-hour inspection. A 100-hour inspection is basically an annual inspection, so if the airplane flies 400 hours in a year, you would face charges for four annual inspections. If it did fly this much though, based on the cost structure we discussed, you'd have generated enough income to cover your expenses.

Tax benefits of any business venture seem to change almost annually. Before you sign any leaseback contract, sit down with an accountant that understands vehicle leasing. If your accountant doesn't understand leasing airplanes—quite a few do not—get a competent vehicle leasing accountant. He can ask all the right questions. If you know an experienced, competent, aircraft owner who can go along with you, that can help too.

Make certain you cover all the worst case scenarios such as the aircraft flying only 10 or 20 hours per month so you can see how costs will vary. Make certain your accountant has closely examined your particular tax status and that when all is said and done, the deal makes sense. No one wants to spent $10,000 to save $5000 in taxes. Tax laws change annually. The most significant tax law change however has been that you, as the leaseback owner, must run your leaseback like a business and not a tax shelter. Certain items are deductible now only in certain situations, so make sure you check with your accountant.

Be certain the aircraft insurance is proper for your particular business venture. If you have $1 million in assets and intend to operate the business as a sole proprietor, an insurance policy that only provides $250,000 in liability protection could be quite a mistake. Be certain that the policy shows the FBO as a named insured and easily points out just what kind of experience the pilots who fly the aircraft must possess.

Lastly, before any agreement is signed, be certain it's been examined backwards and forwards by an attorney. Leases can be negotiated, so if yours doesn't sound good to an attorney, talk about it and suggest changes. If the attorney doesn't understand airplanes, find one who does. Consider contacting the Lawyer-Pilots Bar Association at 500 E St., S.W., Suite 930, Washington, D.C. 20024.

Don't let yourself be dragged into an agreement for leasing an airplane by the smooth lines of the machine or the tax dollars you'll save. Leasing airplanes is a business.

PARTNERSHIPS

If you don't have a head for business, nor the pocketbook to support an airplane on your own, one of the simplest solutions might be to find a partner for the airplane.

When two or more people make an agreement to purchase an airplane that neither could afford on their own (Fig. 4-4) they simply split all the costs by the number of people involved in the deal. Some of these deals have been going on for years with nary a squabble. Others have had a few minor problems that were solved quickly because the partners did one of the smartest things they could have possibly done, *before* they purchased the airplane. They realized that friendships

Fig. 4-4. A fast four-place Commander 114B.

can often be spoiled where a great deal of money is concerned unless some provision is made to deal with the problems that arise when people deal with a subject they are emotionally and financially tied to.

The solution is to produce a written agreement, to spell out how certain kinds of situations will be handled. Too often deals made among friends flow smoothly until a wing tip is crunched in a taxi accident. Because both partners agree to split everything 50/50 a simple plan would be to spread the cost of the repair. Human nature being what it is though, the partner who was not involved in the mishap feels a great deal of resentment toward the idea of having to pay for his partner's negligence. Indeed, partnerships have been destroyed over this sort of problem. If you plan ahead with a solid written agreement, however, you'll find the anxiety level of aircraft ownership can be greatly reduced.

It might sound self-defeating to write an agreement between two friends, because it sounds as if the partners don't trust each other. Certainly, I have known partnerships that were simply run on a handshake and survived intact for years. Perhaps that can work for you, but it's better to be safe than to be unhappy a year down the line with a man or woman who has suddenly started displaying the character of Mr. Hyde.

A better way to look at the written partnership agreement is to consider it like an insurance policy. If anything happens, there's no doubt what the other partners can expect. It's all there in writing and it allows no chance for someone to make an emotional decision that might well jeopardize the aircraft or the friendship. Remember, the written agreement is designed to keep your friendship/partnership intact during moments of stress, like when the aircraft's engine throws a rod because your partner used the throttle like the gas pedal in his 1956 T-Bird.

The written agreement doesn't have to be complicated. It can simply state the names of the partners and the amount of the monthly payment. The idea is to limit the agreement or expand it at will to whatever items all the partners feel comfortable with.

Notice that I did not give all the solutions, but posed some of the questions that should be asked while drawing up the agreement. Situations are too varied to be able to tell what will be right for each person. Sit down with the partners over dinner and hash out the details, but hold the libations until after the agreement is finished.

Twenty steps to a happy partnership agreement

1. The full legal name of each partner.
2. Legal address and telephone number for the partnership.
3. The aircraft registration(s) and serial number(s).
4. Aircraft home airport name.
5. The name of the aircraft insurance company and names of pilots covered by the policy.
6. The amount of money to be sent monthly by check to cover loan payments, aircraft storage (tie-down or hangar), and insurance.
7. The name of the person who will send the respective check to the appropriate place. If payments are due the 15th of each month, then partners might have to submit their respective payments one week prior to the due date.
8. Open a bank account exclusively for the aircraft expenses.
9. The name of the person who will keep the financial books for the aircraft and issue a monthly or quarterly statement of activity for the partners.
10. An agreement regarding where the aircraft maintenance will be performed and which partner will be responsible for overseeing the maintenance.
11. Contingency plans for emergency maintenance when the aircraft is away from the home base. Excessive repair costs should require consent of at least one additional partner.
12. Contingency plans for an incident or accident. Ensure that proper authorities will be notified according to regulations. Determine where responsibility will be placed. It might be easy to point a finger at the pilot in command, but what if another partner landed hard the day before and damaged the landing gear that subsequently collapsed during takeoff the next day?
13. What happens when a partner or partners decide to leave the partnership? How will that person's equity in the aircraft be handled? Too often this issue destroys partnerships because no single person has the money to buy out the departing partners and the aircraft is sold to cover the debt.

14. Determine minimum certificate and experience requirements for all partners and how to handle inclusion of new partners.
15. Determine procedures for acquisition of additional aircraft or replacement of an aircraft. Perhaps there should be a buffer period of 90 days when any dissension can be settled to the satisfaction of all parties.
16. Establish a process for consideration of moving the aircraft to a new home base if storage costs become prohibitive or a partner merely wants the aircraft closer to his home.
17. Establish a policy regarding operation of the aircraft by any nonpartner with adequate flying experience. (The insurance policy will play a major role in this decision.)
18. Determine which partner will be responsible for acquiring insurance and also determine parameters for adequate insurance coverage at the best possible rate.
19. Determine how long a partner may keep the aircraft. Establish a permanent schedule (a large calendar in the aircraft or hanging in a hangar will do) that is accessible to all partners. Establish allowances for vacations, maintenance, and the like.
20. Establish a provision under which a partner can be removed without consent. If one person starts to abuse the equipment, the actions are detrimental to the whole partnership and will be reflected in higher maintenance costs and reduced resale value of the aircraft.

FLYING CLUBS

Another easy way to fly cheaper is by joining a flying club. Clubs are private organizations that bring low-cost flying to the members only, so clubs tend to remain somewhat hidden. The best way to locate the club nearest you is to wander around the local airport and ask questions of the people you meet. "Are there any local flying clubs around? Where do they meet or who could I contact about it?" The best you might come up with is the location of a hangar where the club keeps an airplane. You might have to scout around for who owns the hangar and ask them for a contact person to the club.

The local FBO might be willing to give you the name and address of the club contact person, even though it would at first seem a conflict of interest with the FBO's aircraft rentals. Usually, the club will make a deal with an FBO on the field to purchase fuel and other supplies in a quantity discount agreement. The more the flying club flies, the more money the FBO makes. He might even have some of the club flyers hanging on his bulletin board. The main idea is that flying clubs are a good deal to members who want to fly, as such they usually don't need to advertise.

What is it that makes flying clubs such a darn good deal? Quite simply, the rental rate. If a Cessna 172 costs $50 per hour from the FBO, you might be able to rent a similar aircraft from the club for $40 per hour including the fuel. The club will usually offer similar discounts on various piloting goodies such as books and computers. Of course, the next question is that if these clubs are such great deals, why doesn't everyone join one?

Many pilots do not join clubs because flying clubs operate differently than an FBO. To enter a flying club you must usually be approved by a membership committee, as well as pay an initiation fee, sometimes between $100 and $500. This basically buys a membership in the club, but it doesn't end there. Each month, the club imposes dues upon its members whether they fly or not. Depending upon the aircraft in the club you fly, I've seen dues of as little as $12 to as much as $45 per month. The club could make nearly $600 profit from you over the course of the year, whether you fly or not.

Let's compare FBO and club rates for someone who doesn't fly very often. Renting from an FBO, 50 hours during the year at $50 per hour = $2500. The club rate without the initiation fee, because it is only paid once, is 12 months at $25 per month dues = $300, plus, 50 hours of flying at $40 per hour = $2000, which yields a grand total of $2300.

The club airplane actually cost $46 per hour because you flew so little; toss in the initiation fee to determine first-year costs and that would probably add another $4 or $5 per hour—right where you were with the FBO, $50 per hour.

Fly 200 hours for the year and see the difference: FBO at $50 = $10,000 and club at $40 = $8000. Factor in $1.50 per hour for club dues and realize that the FBO would have cost $10,000, while the club would have cost $8300. You saved $1700, or 42 extra flying hours.

It's your money—give it away or be a careful shopper and fly with it. Flying is expensive, but so is sailing and coin collecting.

5

Some best
first airplanes

WHEN I WAS AT THE airport coffee shop and talked about some of the best first airplanes, people would nod their heads in approval over certain machines and shake their heads in disapproval on others. They also said some of their personal favorites were missing from my list.

I've flown all the aircraft on my list and each has a special place on the list. Some aircraft are absolutely simple to fly while others require extra skill from the pilot. Some are covered in fabric while others encase pilot and passengers in aluminum that is either brightly polished or painted. Most of the aircraft are no longer in production, so the only models you'll find are used, but that does not necessarily mean they are going to be dirt cheap.

An economic recession and restrictions to general aviation flight—due to the controllers' strike and a steadily upward spiral of product liability suits against general aviation manufacturers—sent the cost of new aircraft through the ceiling. This was accompanied by a decrease in the total number of aircraft built. Cessna, Piper, and Beech used to be the big three manufacturers. They now produce only a handful of aircraft per year, with Cessna producing no single engine aircraft in the late 1980s nor since. Piper aircraft has run the line on bankruptcy a few times and as I write this in mid 1993 is still searching for a way to continue producing aircraft. This has been a boon to the used aircraft market because the scarcity of good aircraft pushed prices on available aircraft right through the ceiling. Time was that most aircraft owners watched their aircraft depreciate as time passed. This trend turned around in the late '80s, when owners found their machines worth more than they had originally paid.

For example, in 1978 I purchased a brand new 1978 Cessna 152 for approximately $18,000 to use in a flight training and rental operation near Chicago. When I sold the aircraft two and a half years later with about 850 hours on it, it brought a measly $12,000. In that short time span, my airplane was worth 30 percent less than when it was new. In 1993, that same 152 with about the same amount of time would be worth nearly $21,000. While that might not seem like such a terrific improvement over the $18,000 purchase price, it's a significant improvement over the $12,000 selling price. Other aircraft are experiencing the same kind of appreciation.

Another airplane I owned, a 1968 7ECA Citabria, was purchased in 1974 for $2800. I kept it for three years and flew it more than 600 hours. I sold it for $3500 and thought I'd made quite a deal. That same machine, 15 years later is worth more than $15,000. If there were ever words that I've uttered more than once with individual aircraft that I've owned, it's that I should have kept them. It seemed the aircraft had become too expensive to keep for the amount of time on each or had become a bother. Every time I've sold in the past, I've come to regret my decision.

The question regarding aircraft value might well be how long it will all last. I certainly don't claim to be any expert in predicting the future. I think we can safely say that the prices should remain relatively high for used airplanes because there are so few new ones being produced.

When production of the Cessna 152 ended in 1985, a new 152 sold for more than $30,000. It seems a safe bet that even if Cessna began production of this model again, the price, considering inflation and the few that would likely be produced, would approach $50,000.

I'd buy a used aircraft soon because each year you wait the price seems to increase more than the inflation rate.

USED AIRCRAFT POPULARITY

Chapter 4 discussed the costs that go into paying for an airplane. The highest cost is usually the mortgage. Keep in mind as you consider which airplane to buy, that most aircraft loans run five to seven years. After the mortgage is paid off, costs to operate the aircraft drop significantly. This is why so many people are buying and keeping used aircraft. They'll never be this cheap again.

Another factor that makes these aircraft so popular is they are relatively simple machines. None have retractable landing gear, a variable-pitch propeller, or complicated flight instruments. Mechanically, airplanes are placed in the same "repair class" as those big old Caddies: the older they are and the more options they have means more money spent for repairs.

Stay with a simple airplane until you know more about complex airplanes and related maintenance.

Of the 11 aircraft on the following list, three are fabric-covered taildraggers. To a new pilot, these aircraft—often called conventional-geared aircraft because they lack a nosewheel and use a tailwheel for steering—can seem like quite a handful. Taildraggers require more pilot skill than nosewheel aircraft. Some pilots don't want the bother; they prefer letting the airplane do all the work. Others, however, enjoy the thrill of an airplane that requires them to become one with the airplane; they enjoy an aircraft that asks them to be more than an observer. Some people feel only taildragger pilots are real pilots, but that's a controversy that seems to originate mostly from taildragger pilots.

Another item that makes these older aircraft slightly different from newer aircraft is that they are covered in fabric, something that seems archaic today. Years ago, most aircraft were nothing more than a welded fuselage of steel tubing. Some were just a wooden frame with a high-grade cotton cloth stretched across the airframe and wings. This cloth was then coated with a dope paint (ever build model airplanes?) and sent on its way. It was not unusual to use doped fabric in many kinds of aircraft. The former queen of the commercial skies, the Douglas DC-3, used fabric-covered control surfaces, such as the elevator and ailerons, while the rest of the aircraft was covered in metal.

Some people think that a fabric aircraft is cheap or unsafe. Nothing could be further from the truth. Fabric in good shape can be extremely strong per square inch. It can be slit with a knife or punctured with a sharp object, but it certainly won't rust or corrode like steel. Learn about fabric or metal before you buy to be an educated consumer.

Before I begin throwing various kinds of aircraft performance data at you, let's make sure you understand the terms I'm talking about:

- *Wingspan* is the distance measured from wing tip to wing tip.
- *Length* is from prop spinner to tail.
- *Height* is to the top of the rudder.
- *Empty weight* is measured with radios installed, unusable fuel and no people on board.
- *Maximum weight* is the heaviest the aircraft can legally be for takeoff.
- *Useful load* is essentially baggage, people, and fuel.
- *Cruise speed* is the speed at which you would normally fly the airplane.
- *Service ceiling* is how high the airplane will fly under normal conditions before the engine cannot produce enough horsepower, because of the thin air, to keep the airplane aloft.
- *Takeoff and landing distance* are self-explanatory.

Average prices quoted are in 1989 dollars and vary depending on the year of manufacture and the equipment installed in the aircraft.

CESSNA 150/152

The Cessna 150 was first produced in 1958 and is just about the most popular trainer around. The earliest 150s sported a 100-hp Continental engine and cruised around 105 mph as they burned about five to six gallons of fuel per hour. The aircraft is small, inside and out. The trailing edge of its high wing stands only about five-and-a-half feet above ground level and has often caused a few bruised heads and noses on people walking into them.

The aircraft is one of the lightest on the controls of any model in the list. While the cabin tends to be tight for two large people, the aircraft can be equipped with any radio or instrument available. The maintenance on the 150 is relatively inexpensive due to its metal construction and basic systems.

The 152 (Fig. 5-1) is essentially the same aircraft as the 150, except the up-

Fig. 5-1. Cessna 152.

graded 110-hp Lycoming engine. Many oil companies stopped producing 80 octane aviation fuel in the mid-'70s, which the 150 ran on beautifully. They produced only 100 octane, which the 150 didn't enjoy. The 100 octane resulted in many cases of burned valves in 150s. The new 152 is designed to run off 100-octane fuel with more durable exhaust valves. However, the spark plugs on this engine tend to foul rather quickly, based on my experience. The 152 can be difficult to start when the temperature is around 35 degrees or lower, but it's a fine flying aircraft.

Cessna 152

Wingspan   33 ft 2 in
Length   24 ft 1 in
Height   8 ft 6 in
Engine   Lycoming O-235 (110 hp)
Fuel capacity   22 gal standard, 39 gal long range tanks
Empty weight   1104 lbs
Maximum weight   1670 lbs
Useful load   571 lbs
Seats 2
Maximum speed   125 mph (110 kts)
Normal cruise   119 mph (107 kts)
Service ceiling   14,700 ft
Takeoff distance   725 ft
Landing distance   475 ft
Average 1993 price of Cessna 150 $6000 to $18,000
Average 1993 price of Cessna 152 $12,000 to $28,000

CESSNA 172

One of my favorite aircraft to fly, the Cessna 172 (Fig. 5-2) combines the smoothness and control lightness of the 150/152 but adds the comfort of a larger four-place cabin and the more powerful 160-hp Lycoming that burns about eight gallons of fuel per hour. Many new pilots make the 172 their first buy because most are equipped for IFR flight. The cost of an instrument rating as well as the added utility of the rating can also be factored into the aircraft cost. Personally, I can't think of one unpleasant characteristic of the 172.

Cessna Aircraft Co.

Fig. 5-2. Cessna 172.

Like the 150/152, the 172s have been produced for 30 years, which means you can find one with just about any kind of equipment. If you decide to buy, there is a special broker who specializes in 172s (1-800-SKYHAWK).

Cessna 172

Wingspan   35 ft 10 in
Length   26 ft 11 in
Height   8 ft 9½ in
Engine   O-320 Lycoming (160 hp)
Fuel capacity   43 gal
Empty weight   1414 lbs
Maximum weight   2407 lbs
Useful load   993 lbs
Seats 4
Baggage weight   120 lbs
Maximum speed   141 mph (123 kts)

Normal cruise   138 mph (120 kts)
Service ceiling   13,000 ft
Takeoff distance   890 feet
Landing distance   540 feet
Average 1993 price of Cessna 172 $12,000 to $60,000

PIPER CHEROKEE

The earliest of the modern-day Piper trainers was the Cherokee (Fig. 5-3). It was a low-wing airplane with flying characteristics so simple that the airplane was almost impossible to stall. Training in the early Cherokees, I would try to induce a stall in the fat symmetrical wing. Usually I was too slow to make the aircraft demonstrate that standard break or dropping sensation common to many high-wing types. Most of the time I was able only to keep the aircraft mushing along—flying, but not really flying. I was always under control: terrific thing for a new pilot.

Fig. 5-3. Piper Cherokee.

Piper Aircraft Corp.

I believe that certainly an airplane with a low wing placed so the pilot can see forward allows a much better view of things than the high wing offers. When you're aloft you always feel like you're flying because you can easily see from horizon to horizon as well as the clouds and sky above. It's a more open feeling. Many pilots who began flying in high-wing aircraft tend to remain in those airplanes out of habit and comfort. The same holds true for pilots who take their training in low-wing machines. Although much of my flying and teaching experience is in high-wing machines, it's difficult to choose one over a low wing that appears to spread the earth and route of flight before me totally unencumbered.

Early Cherokees ran with a 150-hp Lycoming and later models were termed the Cherokee 180 (180-hp) and the Cherokee 235 (235-hp). The 235 contained a six-cylinder engine and obviously burned the most fuel.

In the mid-1970s, Piper changed the names of these singles to reflect a different marketing scheme. The Cherokee 140 with an upgraded engine and interior became the Archer. The 180 Cherokee became the Warrior and the six-cylinder Cherokee 235 emerged as the Dakota.

Piper airplanes are roomy with wide instrument panels that allow plenty of space for the installation of most any radio you'd care to buy. The baggage room on most Cherokees is usually quite ample.

Piper Archer

Length   23 ft 8 in
Height   7 ft 3 in
Cabin width   41.5 in
Cabin height   49 in
Engine   Lycoming O-320, 160 hp
Fuel capacity   48 gal
Empty weight   1535 lbs
Maximum weight   2325 lbs
Seats 4
Maximum speed   146 mph (127 kts)
Normal cruise   141 mph (123 kts)
Rate of climb   710 fpm
Takeoff distance   975 ft
Landing distance   595 ft
Average 1993 price for Cherokee 140 $12,000 to $18,000
Average 1993 price for an Archer $32,000 to $65,000

PIPER TOMAHAWK

Piper Aircraft Corporation designed a new trainer, the Tomahawk, in the late 1970s (Fig. 5-4). It is smaller than a Cherokee, with seats for two people. It incorporated a new T-tail design that put the horizontal stabilizer out of the normal airflow to smooth the ride. It also incorporated an even larger amount of window area to give the student a better view of the world around him. While not terribly fast, nor able to carry a large amount of baggage, the Tomahawk certainly was economical with its O-235 Lycoming engine similar to that used in the Cessna 152. While thousands of the airplanes were manufactured, they never did quite become the staple of the flight schools the way the Cessna 152 or the Cherokee line did. This is reflected in their prices on the used aircraft market.

Fig. 5-4. Piper Tomahawk.

Tomahawk

Length   23 ft
Height   9 ft 6 in
Cabin width   42 in
Cabin height   50.5 in
Engine   Lycoming O-235, 112 hp
Fuel capacity   32 gal
Empty weight   1109 lbs
Maximum weight   1670 lbs
Useful load   561 lbs
Seats 2
Maximum speed   125 mph (109 kts)
Normal cruise speed   124 mph (108 kts)
Rate of climb   718 fpm
Service ceiling   13,000 ft
Takeoff distance   820 ft
Landing distance   707 ft
Average 1993 price of Tomahawk $8500 to $13,000

BEECHCRAFT SKIPPER

It is difficult to look at a Beechcraft Skipper (Fig. 5-5) and tell it apart from its T-tailed neighbor, the Piper Tomahawk. This machine also incorporates plenty of glass area for excellent visibility and also uses the trusty O-235 Lycoming engine. The main difference used to market the Beech Skipper against the Piper Tomahawk was careful attention to detail by Beech: cloth or crushed cloth seats instead of vinyl. Very often, the Beech instrument panel was more attractively presented than Piper's. There was a certain amount of truth in the Beech marketing plan against the Tomahawk because the empty weight of the Skipper was nearly 500 pounds more than the Toma-

Beech Aircraft Co.

Fig. 5-5. Beechcraft Skipper.

hawk, probably indicating a stronger fuselage and wings. The Beechcraft also carried considerably more fuel for better range than the Piper.

The Skipper had a higher price tag than its competitor. Otherwise the machines are virtually the same. One advantage of the Skipper was the wide landing gear that made crosswind landings easier for novice pilots.

Beechcraft Skipper

Wingspan   32 ft 9 in
Length   25 ft 9 in
Height   8 ft 3 in
Cabin width   44 in
Fuel capacity   57 gal
Empty weight   1505 lbs
Maximum weight   2450 lbs
Useful load   950 lbs
Seats 2
Maximum speed   141 mph (123 kts)
Normal cruise   136 mph (119 kts)
Rate of climb   792 fpm
Service ceiling   12,600 ft
Takeoff distance   1130 ft
Landing distance   703 ft
Average 1993 price for Beechcraft Skipper $15,500 to $20,000

BELLANCA CITABRIA

Citabria came from the word airbatic spelled backward. (I have no idea how they figured that one out.) The first of these fabric covered taildraggers was put on the market in 1964 and used a 100-hp engine. Somewhat underpowered, the airplane was later improved with, again, the O-235 Lycoming developing 115 hp (Fig. 5-6).

Fig. 5-6. Bellanca Citabria.

The design was hardly classically aerodynamic, but it could carry two people and baggage comfortably on a long trip or be used for fun flying. This plane has front-and-back tandem seating instead of side-by-side and has a stick for elevator and aileron control instead of a control wheel.

The airplane was aerobatically certified. It is perfectly suited to teach a pilot how to perform a loop, a roll, a split S, or a good spin.

Many Citabrias have a greenhouse roof in the ceiling that let a pilot see what is going on above the airplane—or to see while inverted. All Citabrias have a welded steel-tube fuselage covered with Ceconite fabric. The wing spars and ribs are made of wood.

Fabric-covered wooden-wing airplanes pose a different set of problems for the used aircraft buyer, such as the need to check for cracks or rot, but they can be as sound as a metallized machine.

Several different models of the Citabria were built. The following data are for the 115-hp 7ECA. It was the only 115-hp version built. Next came a 7KCAB with a 150-hp fuel-injected engine and a complete inverted fuel and oil system. The 7GCAA and 7GCBC are non-fuel-injected 150-hp versions of the Citabria.

7ECA Citabria

Length   22 ft 7 in
Height   7 ft 8 in

Wingspan   33 ft 5 in
Engine   O-235 Lycoming
Fuel capacity   35 gal
Empty weight   1060 lbs
Maximum weight   1650 lbs
Useful load   590 lbs
Seats 2
Maximum speed   129 mph (109 kts)
Normal cruise speed   124 mph (105 kts)
Range   694 statute miles
Service ceiling   12,000 ft
Takeoff distance   450 ft
Landing distance   890 ft
Average 1993 price for a 7ECA Citabria $14,000 to $21,000

GRUMMAN TRAINERS

American Aviation introduced the Yankee in the late 1960s, which was subsequently produced by Grumman American Corporation (Fig. 5-7).

The slick little airplane has a sliding canopy, so you enter by stepping over and in, just like a fighter plane. The skin is not aluminum or fabric like most light aircraft. The skin on Grummans is glued composite material that requires few rivets and screws. This makes the outer skin a smooth finish, almost like a plastic model.

Later models included the Tiger, Cheetah, and the twin-engine Cougar. The Yankee and the Trainer used the 115-hp engine. The nosewheel castered and the pilot learned to steer by using differential braking.

Fig. 5-7. Grumman Trainer.

AA-1 Yankee and Trainer

Length 19 ft 2 in
Height 7 ft 6 in
Wingspan 24 ft 6 in
Engine O-235 Lycoming (115 hp)
Empty weight 1066 lbs
Maximum weight 1600 lbs
Useful load 534 lbs
Seats 2
Fuel capacity 24 gal
Range 300 statute miles
Service ceiling 11,500 ft
Takeoff distance 890 ft
Landing distance 425 ft
Average 1993 price of Grumman Trainer $9000 to $14,000

PIPER CUB AND SUPER CUB

If ever there was an airplane that proved to be a public relations windfall for general aviation, it was the Piper Cub (Fig. 5-8). Whenever a novice talks about small airplanes, Piper Cub is used, no matter what it might actually be.

Fig. 5-8. Piper Cub.

Glenn Chatfield

Originally produced in the early 1940s, the Cub combines simplicity of design with the need for a pilot to fly this airplane. The Cub is a tailwheel aircraft and displays different characteristics on landing, takeoff, and taxi, than most tricycle-trained students are used to.

In the air, the aircraft is relatively heavy on the controls but gives the pilot a solid feel, a solid belief in the stability of what they are flying. Cubs are easy to spot, too. Most of them are painted yellow.

Cubs are so versatile and display such a mild-mannered attitude in slow flight that they've been used for everything from police patrol to search and rescue to a tow plane for gliders and banners. Contrary to some opinions, Cubs are fun to fly despite their control heaviness. I can think of no greater joy than the time I flew with a friend in his Cub with the clamshell door removed. It was a warm summer afternoon and we squandered an hour or so flying over farm fields at about 200 feet, waving to the people on the ground. They waved back. We cruised at 85 mph.

Original Piper Cubs arrived from the factory with 65-hp engines that sniffed fuel more than drank it. Later Cubs used engines through 95-hp. The Cub evolved into the Super Cub when Piper began installing 150-hp engines in the aircraft, in addition to building a bigger airframe.

Piper Cub

Length   22 ft 4 in
Wingspan   35 ft 2.5 in
Engine   A-65 Continental (65 hp)
Empty weight   680 lbs
Maximum weight   1220 lbs
Useful load   540 lbs
Seats 2
Range   179 statute miles
Service ceiling   11,500 ft
Takeoff distance   200 ft
Landing distance   350 ft
Average 1989 price of Piper Cub $9000 to $13,000
Price of new Piper Super Cub in 1993 $55,000

TAYLORCRAFT

When I was a kid, I thought it was a pretty simple matter to separate the various airplane models from one another at the airport. There was one pesky model though, the Taylorcraft, or T-Craft as it's called, that I always had trouble with. At times I couldn't be sure whether or not I was looking at an Aeronca Champ or the T-Craft. Let's say the T-Craft holds almost as high a place in the inexpensive airplane world as does the Cub.

The T-Craft is a combination of wood, fabric, and steel. The cabin is small, as is the available space to haul baggage. The airplane is slow but burns only about four gallons of fuel per hour.

The T-Craft, like other aircraft designed in the '40s, was developed to introduce inexpensive transportation over short distances to as many pilots as possible. This is not the machine to buy if you intend to fly 300 miles one way to the cabin every weekend. This is a fun airplane to fly an hour or two every chance you get, just enjoying the fact that you can fly.

This is the airplane you can take to another airport for a quick cup of coffee or lunch with some of the airport people you meet—who are also plane crazy. T-Craft are not as plentiful as they once were, but a good one is worth the search.

Taylorcraft

Length   22 ft 6 in
Height   6 ft 6 in
Wingspan   36 ft
Engine   O-200 Continental (100 hp); older versions used 40, 50, and 65 hp engines
Maximum speed   127 mph (110 kts) new version
Cruise speed   123 mph (104 kts)
Empty weight   900 lbs
Maximum weight   1500 lbs
Fuel capacity   24 gal
Seats 2
Range   450 statute miles
Service ceiling   18,000 ft
Takeoff distance   300 ft
Landing distance   375 ft
1993 price of used Taylorcraft $6500 to $13,000

CESSNA 120 & 140

If you've traveled around to most any general aviation airport, you've almost certainly seen a Cessna 150 or 152. Many new aviators believe these were the first of the Cessna trainers and are even surprised to hear that another famous Cessna preceded, in fact spawned, the 150 line. It was the Cessna 120/140 series (Fig. 5-9) and over the course of years, Cessna produced 7,500 of them. The 120/140s were a new generation of light aircraft, encompassing a metallized fuselage and, on early models, fabric wings. Later models were completely metallized. Like the T-Craft, the Cessna 120/140 series was not terribly fast, nor would it carry a tremendous amount of baggage.

The original 120/140s are conventional taildragger airplanes. They used the Continental 85-hp engine that would burn about 4.5 gallons per hour, making them pretty easy on the pocketbook. Even today, with a fairly large number of the old machines still flying, the insurance rates and general operating costs are so low that a 120/140 can easily fit into most any pilot's budget. The only basic difference between the Cessna 120 and the later 140 is flaps on the 140. Most pilots found the flaps to be useful only for slowing down. Many pilots have converted the 140 into a 120 by removing the flaps.

A Cub or a T-Craft with a great deal of cockpit instrumentation is an unusual sight. Many of the Cessna 120/140 aircraft that visit the EAA air show at Oshkosh

Glenn Chatfield

Fig. 5-9. Cessna 140.

each year carry a full array of flight instruments and radios, often enough to fly IFR. Hard IFR in a slow Cessna 120 can cause a great deal of grief to the ATC system because of the speed. A fully IFR C-120 could certainly be used to fly on those days when a short climb through an 800-foot ceiling would put you on top at 2200 feet into the bright sunshine for an afternoon's fun. And it would be comforting to know that if the weather should drop below VFR, you could make an IFR approach. This would be better than attempting the standard—and dangerous—scud running.

Cessna 120/140

Length   24 ft 1 in
Height   7 ft 1 in
Wingspan   32 ft 10 in
Engine   Continental 85 hp through Lycoming O-235 and O-290
Maximum weight   1450 lbs
Empty weight   850 lbs
Useful load   600 lbs
Fuel capacity   42 gal
Seats 2
Range   550 miles
Maximum speed   125 mph (106 kts)
Normal cruise   120 mph (102 kts)
Service ceiling   18,000 ft
Takeoff distance   250 ft
Landing distance   400 ft
Average 1993 price of Cessna 120/140 $7800 to $12,500

BEECHCRAFT SUNDOWNER

The Beechcraft Sundowner (Fig. 5-10) was the Beech Musketeer in the 1960s. It evolved over the years into a comfortable, quiet machine. Some of the airplane's critics compare it to the Cessna 172, which is a bit faster than the somewhat boxy looking Sundowner. The price for a Sundowner is in keeping with the Beech tradition of being the most expensive of its class. The instrument panel and the cabin of the Sundowner are very big. The panel is wide enough that, even after the standard complement of IFR radios has been installed, plenty of extra space remains for future additions.

Fig. 5-10. Beech Sundowner.

Baggage space abounds in the Sundowner and with the low wing, entry and exit is easy through the wide door on either side of the cabin. The pleasurable part of owning a Sundowner is flying. While the aircraft does have the heavier feel of a larger aircraft, it is pure joy once airborne. The interconnecting springs between the ailerons and rudder give the pilot a chance to cruise and turn with very little effort.

The extra factory soundproofing gives the pilot and passengers something few other aircraft in this category can find—quiet. It is actually possible to fly a Sundowner at cruise power and converse with someone in the next seat without yelling, something that is almost impossible in the competing Cessna 172 or Piper Cherokee. Prices for used Sundowners tend to be high, but it could well be worth checking out before you buy anything else.

Beech Sundowner

Wingspan   30 ft
Length   24 ft
Height   6 ft 11 in

Engine   180 hp Lycoming
Fuel capacity   58 gal
Range   650 miles
Empty weight   1103 lbs
Maximum weight   1675 lbs
Useful load   577 lbs
Seats 4
Maximum speed   122 mph (106 kts)
Normal cruise   121 mph (105 kts)
Rate of climb   720 fpm
Service ceiling   12,900 ft
Takeoff distance   780 ft
Landing distance   670 ft
1993 price of Beechcraft Sundowner $17,000 to $39,000

Certainly, buying the first airplane is a decision you should not take lightly, nor make too quickly. Different equipment in the different models outlined here can change the price of the aircraft. Owning an airplane can be a great experience, or a nightmare if you don't do your homework.

Whether you're an experienced pilot or not, make certain any aircraft you decide to buy has a rider. This rider will allow you to take the aircraft to a qualified mechanic of your choice for a prepurchase inspection and a renegotiation of the original terms of the deal if that inspection discovers problems with the aircraft.

6

Advanced ratings

ONCE YOU'VE PASSED the private pilot checkride, the furthest thing from your mind is more training. But, as a wise old instructor once told me: "Your private license is only a license to learn." That phrase rings with a certain amount of truth. The first 50 hours after the flight test should be devoted to fun. Make it a point to take trips to hone your cross-country skills. You'll find there really was a purpose to all the information you learned in ground school.

As you fly those cross-countries, don't just fly a straight line from point A to point B. Make the trip a challenge; try various techniques. Make one leg of the journey VOR navigation and another leg dead reckoning while a third leg could be pilotage. Throw a curve ball and imagine that 10 miles from a destination the weather goes below VFR. What would you do? Then do it.

You need not fly long distances to perfect cross-country techniques. Figure 6-1 is a good example of a zigzag cross-country route. It is a varied route that eventually puts you to the test with different navigation and radio techniques.

And that's really what this is all about. You'll never feel comfortable flying cross-country routes if your skill levels are poor and you'll never improve without a challenge.

Set aside time each week to fly in the local traffic pattern. If you have the time, leave the pattern and fly around the local area. Make it fun, but try to maintain the challenge. Try to see how much farther you can roam each time you fly locally without looking at the sectional. Learn the landmarks farther and farther from your home plate.

Let me qualify the last paragraph with one point. In recent years the FAA has stepped up its enforcement action against pilots that violate protected airspace. This is more than flying through a restricted area or, heaven forbid, a prohibited area. I'm referring to the airspace around airport radar service areas (ARSAs changed to Class C airspace, in September 1993) and terminal control areas (TCAs changed to Class B airspace, in September 1993). In 1986, a single-engine Piper entered a portion of the Los Angeles TCA and collided with a DC-9 on final approach to LAX. All aboard both aircraft were killed.

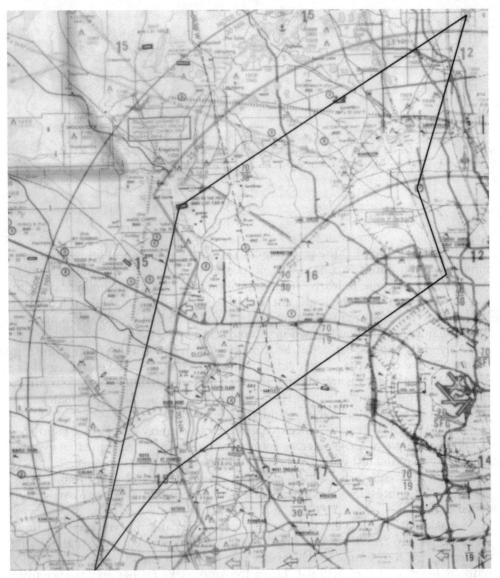

Fig. 6-1. Sample navigation chart for short cross-country flight.

The enforcement repercussions of that accident have been far reaching, with license suspensions now common. The regulations require that you know where you are. Airspace has certainly become more complex in the past 10 years. If you live near an airport with a TCA or an ARSA, consult or fly with a local flight instructor to be sure you know what you are doing.

One FAA regulation requires flying with an instructor every two years to redemonstrate your flying skills. The flight reviews are not tests, so you can't flunk.

But the instructor must make a notation in your logbook that the ride was completed.

I have never failed anyone on a flight review, but I have refused to sign the logbook of a pilot I thought really had no business flying an airplane alone. Usually an extra hour of dual instruction after the first ride was enough to put the person back on track.

Let's assume you have flown 50 or 100 hours since acquiring your private license and you've asked: "Is that all there is?" The answer is an unqualified "no," there's still plenty you can learn about flying and aviation if you want to take the time.

What about advanced ratings? What is available? Where? How much will they cost? Are they worth it?

You can add several ratings to a private pilot certificate or you might earn another certificate or you might do both.

There is no such thing as a useless rating or license. When I moved past my private license to commercial, instrument, multiengine and flight instructor levels, I never envisioned teaching anyone anything. I think the learning just became a habit. The more I learned the more I wanted to know about flying. The more I wanted to smooth and perfect my skills, and understand the little quirks I saw in airplanes and flying. Like the wonderfully childlike and inquisitive robot in the film *Short Circuit*, I wanted more input.

Where do you start? What should you try next? This really is a question only you can answer after you sit down with paper and pencil and decide why you are trying to pursue flying past your private.

If you fly for business, perhaps you've been grounded more than a few times because of weather. If that's the case and the airplane you fly is instrument equipped, an instrument rating could be the answer. If your tastes have changed in your present profession and you think a career in aviation might be more interesting, a commercial certificate would be your next goal. The commercial certificate will also give you a great deal of experience flying the aircraft through new experience building maneuvers.

If you enjoy the thrill of aviation and don't have a particular need for either of these license additions, perhaps a rating that is just plain fun might be the better answer. Consider a glider or hot air balloon rating. If you want a real challenge and have the time and funds available, why not add a rotorcraft-helicopter or gyrocopter to your private certificate?

THE INSTRUMENT RATING

As the author, I have the privilege to give personal recommendations about ratings. Considering the state of aviation today and the current philosophy of the FAA regarding pilot errors, it's very important that you, as pilot-in-command (PIC), know where you are. You also need to know how to cope with various types of airspace. Nothing teaches a pilot more about navigation, communications and weather than an instrument rating.

FAR Part 61 says you must have accumulated at least 125 hours of flight time to be eligible for an instrument rating. Fifty of these hours must be pilot-in-command on cross-country flying. This means total flying time when you take your instrument checkride, so you can begin working toward the instrument rating much earlier.

Another important point. When moving on to a new rating you will come across rules that must be followed if your flight time is to be properly credited. For the instrument rating, each cross-country flight must include a landing at a point at least 50 nautical miles from the starting point. You're on your honor here because you are not required to bring your logbook in to the destination and have it signed. Just make certain you keep a good up-to-date record in the logbook of where and when you fly. Keep in mind, too, that the rule says 50 nautical miles, not statute miles.

The instrument rating is going to cover several different subjects, but normally begins with basic instrument work. Your time is devoted to learning more about how the basic flight instruments operate, what the problems can be with those instruments, and what the pilot can do to solve them. You'll learn about weather. Not just VFR and IFR, but what produces the weather each day: the fronts, the thunderstorms, the ice, the snow, the fog.

You'll learn how to read the charts and the reports to determine whether the flight you're trying to make is safe or not. An instrument rating allows you to fly in conditions less than 1000-foot cloud ceiling and 3 miles visibility, but it doesn't allow you to fly in just any kind of weather.

You'll learn more about navigation because you'll be flying only by reference to flight instruments and radio navigation gear. That means learning to fly on VOR airways and fly to NDB stations. It means learning to mentally visualize where you are as you fly. You'll better understand the meaning of ATC and communications that allow you to operate safely in the system.

Instrument landing charts will become second nature. The most important factor involved in instrument flying however, is the rating's ability to demand more of you. You must have a good feel for the airplane and what it takes to keep it flying with only a minimal effort. No longer are you going to be concentrating primarily on flying. Rather, flying will need to become almost second nature so you can perform the required navigation and communication jobs necessary to fly in the IFR system.

While this might sound like a great deal of work, it's a kind of work filled with excitement as one block of instruction builds upon another. You don't have to wait until the end of the course to see the progress. You'll see the extent of the learning when you walk out in the backyard some summer afternoon and feel a breeze increasing from the southwest. You'll know that the dark area to the west means an area of prefrontal rain with possible thunderstorms approaching. And you'll know what to expect after the front passes.

Little by little your knowledge will build until you realize you do have a solid footing upon weather and weather predictions.

Flying by total reference to instruments is something you experienced as a student although what you saw as a student was very rudimentary. In fact, the exposure was designed to give you a chance to get away from bad weather if you unwittingly wandered into it, nothing more. For the instrument rating, your instrument flying techniques will be honed to such a level that you'll be able to fly without glancing out the window. You will cope with the emergencies of flying when various pieces of equipment, such as radios, fail.

If you thought you knew how to read a VFR sectional, you haven't seen anything yet because IFR charts look nothing like VFR charts.

You'll learn how to communicate, talking more to air traffic control as well as the specialists at the Flight Service Station (FSS).

An instrument rating is indeed work. However, there is nothing like the feeling of making all the numbers and instruments come together on some dark and drizzly evening as you descend through the clouds on an instrument approach to find the runway approach light system pointing the way to your runway. It's sure better than sitting in a motel in Kansas City waiting for the fog and rain to pass. You'll quickly realize how much utility you've added to your license and your airplane with this rating. Estimated cost in 1993 is $3500.

THE COMMERCIAL CERTIFICATE

If you intend to make money flying, which could be something as simple as towing banners or as a corporate pilot or a flight instructor, you must have a commercial pilot certificate. If you happen to hold a free balloon rating, picking up the commercial certificate for ballooning will give you the authority to act as a flight instructor for balloons. The strictest definition of FAR Part 61.139 says the holder of a commercial pilot certificate may "act as pilot in command of an aircraft carrying persons or property for compensation or hire."

To be eligible for a commercial pilot certificate, you must have a total of 250 hours flying time of which at least 100 hours must be pilot-in-command time. The remainder might be spread out among dual instruction and simulator time. One other major requirement to obtain an unrestricted commercial certificate is an instrument rating. And all the time you put in toward an instrument rating counts toward the commercial. Commercial pilot training will include 10 hours of dual instruction in a complex aircraft (Fig. 6-2). A complex aircraft includes retractable landing gear, variable-pitch propeller, and operational wing flaps.

What's so great about the commercial training? Quite simply, the commercial rating is going to give you more knowledge about flying the aircraft. You'll be learning and practicing additional precision maneuvers such as accuracy landings, ground reference maneuvers, chandelles, and lazy 8s, as well as more intensive instruction in aerodynamics, weight and balance, and commercial flying regulations. If you're going to be in situations where you are responsible for the safety of other people or their property, the FAA wants to make sure you know more about what really keeps an airplane in the air and what brings it down. You must be proficient at crosswind landings. Merely being able to get the airplane into the air and back without hurting yourself or the airplane is not good enough.

You'll learn more about weather and how to fly the aircraft with more precision. Imagine being on a downwind leg, opposite the end of the runway when the

Fig. 6-2. Beech Sierra often flown to satisfy the complex aircraft requirement for a commercial certificate.

instructor requests that the wheels touch down on the big line on the runway. You, with training, will be able to maneuver with enough precision to accomplish this.

You'll also learn about pushing airplanes to the extremes of their flight envelopes, OK, perhaps not exactly to test pilot standards, but certainly flying between the slowest speeds and the fastest. You'll fly maneuvers with reference to the ground while at the same time maintaining a constant airspeed or heading by only occasionally looking inside the cockpit.

The point of the commercial certificate is to make you feel like "one with the airplane." When you do begin flying customers or freight, you'll feel comfortable and easily cope with small problems.

One of my first jobs with my brand new commercial certificate was towing banners with a 150-hp Citabria around the Chicago area. A banner was attached to nylon webbing that stretched for about 50 to 60 feet. The banner could not be attached to the aircraft on the ground. It was necessary to depart the airport, arrive over the banner pickup area, and extend a large hook on a rope behind the airplane. The aircraft descended to around 10 feet off the ground in level flight at maximum airspeed. The cord holding the banner was stretched between two 15-foot poles and I would fly between them, hook up the banner, and fly away to the "target."

It was an incredible experience. For two years I learned how to control an airplane in slow flight accurately, and how to judge distance and heights better than I ever had before, as well as the intricacies of planning fuel loads with high outside air temperatures. Weather was always a problem because I was often over the target for most of the day. I learned to cope with many ATC problems and in general felt so confident about the experience gained from banner towing that I easily moved into flying a light twin-engine Cessna for a small company.

A commercial license can be combined with training for an instrument rating, or the two can be kept separate. The choice is yours.

Estimated cost in 1993 is between $4000 and $8000, depending upon how much flying time you have logged when you start training.

THE MULTIENGINE RATING

When I began my training toward a multiengine rating, I had just received my commercial license. For me, the commercial had been a long and dusty experience, but certainly worth the work. Now I wanted to do something fun, something requiring only a limited amount of work. I chose the multiengine rating because I decided it couldn't be too much tougher to fly an airplane with two engines than to fly with one. I was surprised at what a challenge it was to fly a Cessna 310 (Fig. 6-3).

Fig. 6-3. A Cessna 310, which is often used for multiengine training.

The 310 was bigger and faster than anything I had ever ridden in before, much less flown. There is no minimum number of flight hours required to be eligible for a multiengine rating. In fact, if you could afford it, you could choose to train for the private pilot license in a twin-engine airplane. The reason it's not too common is that most twin-engine airplanes are rather complex when compared to a Piper Warrior or Cessna 172. The more complex the aircraft, the more training it will take to learn to fly.

Multiengine training is typically in a Cessna 310, Piper Seneca, Seminole, Aztec, or Beechcraft Duchess or Baron. Be prepared for a wild ride the first few times out if your total time is low. That first ride in a twin will find you in command of an airplane moving at 180 mph during climb while your brain is still trying to figure out where the boost pump switches and the landing gear lever are located. Luckily, you'll have a flight instructor with you all the way. Normally no solo time is allotted for the multiengine rating.

Initial hours in the twin will be getting used to the systems and the difference between a twin and a single. This means larger and more complex fuel systems, faster speeds for everything, more complex weight and balance computations, and learning to think farther ahead of the airplane. In the Cessna 150, I could turn into the traffic pattern at cruise airspeed and still slow down properly before landing. The first time in the 310 I slowed up early, about two miles out, and found I was still behind the airplane. The airplane didn't want to slow down, so my brain had to speed up.

After I began making passable landings, my instructor started showing me what makes most twin-engine aircraft different from single-engine aircraft. I received demonstrations of what happens to twins when one engine stops running. I learned what the phrase "in command" really meant in a light twin, with demos of engine failures on takeoff, at cruise, and in the landing pattern. Twin-engine airplanes are demanding aircraft to fly, but they can be a lot of fun.

After only about 13 or 14 hours, I felt pretty confident and could fly the maneuvers well enough to pass the checkride. I did this with about 270 hours total flight time. Estimated 1993 cost for a multiengine rating is approximately $2500 depending upon the training aircraft and your abilities although as with all ratings, a bargain can often be found with a little homework.

THE ATP CERTIFICATE

If you consider pilot licenses in the same order as an educational hierarchy, with a private being a high school diploma, the commercial certificate would be equivalent to a college bachelor's degree. At the furthest end of the pilot license spectrum is the airline transport pilot (ATP) certificate, which would be equivalent to a doctorate. According to the Aircraft Owners and Pilots Association (AOPA) there are about 692,000 active licensed pilots in the United States. Of this number, about 112,000 hold an ATP certificate, so you can see it is a relatively small group of pilots indeed. The vast majority of ATPs are airline and corporate pilots flying large jets and smaller turboprops. The remainder are pilots—some just like yourself—picking up the certificate to qualify for a more lucrative future job. Some pilots merely want to join the small group of pilots labeled as some of the best in the country.

The oral exam for the ATP is not that tough. But while the checkride is essentially like another instrument rating checkride, real differences are found in the tolerances. On a private pilot checkride, you're allowed to slip off altitude by 100 feet and heading by 10 degrees. When I took my ATP checkride I asked what the examiner expected in the way of altitude, heading, and tracking of VOR radials or NDB headings. "Your tolerance is zero," he said. It was probably the best flight check I ever flew. When I was finished and had been told that I'd passed, I truly felt as though I'd accomplished something spectacular (Fig. 6-4).

THE CERTIFIED
FLIGHT INSTRUCTOR (CFI)

After earning my commercial certificate and multiengine rating I felt confident in the machines I was flying, but I still didn't have the grasp of aerodynamics and

Fig. 6-4. Piper Seneca can be used for multiengine ATP training.

general flying techniques I wanted. I never intended to teach flying. Training began because I wanted to know more about flying.

The instructor explained exactly what I was going to learn and just what would be expected on the flight test. The description of the oral exam was simple: the examiner can ask anything about flying—anything! The initial phase of flight instructor training began with a test of my general flying abilities. This also served as a check to be certain I did hold the prerequisite commercial license and instrument rating. Next we talked about the type of instructor certificate I was seeking, essentially a certified flight instructor-airplane, single-engine land. Picking up this rating would allow me to give the instruction necessary to sign a student off to take private or commercial pilots checkrides. It also gave me the authority to proctor the written exams necessary for those tests.

The FAA does not explain much about how to teach. This is where your practical training comes in.

Sure you know how to land the airplane and you can perform a respectable turn around a point. But, could you teach someone else how to perform these maneuvers? That's the trick. Oh yes, you must also learn to fly the airplane from the right seat, which for some people can be a bit tougher than it sounds. A private or commercial pilot scans the horizon from a particular place in the cockpit to pick up visual cues. Now the perspective changes, but you'll catch on with practice.

In addition to a mastery of flight maneuvers, aircraft stability, and weather, a flight instructor applicant must be able to demonstrate a feel for the preparation of lesson plans. Teaching also includes the ability to satisfactorily critique a student's performance (Fig. 6-5). That doesn't mean merely telling a student when he or she has made a mistake, but also what they can do to improve their technique. Recall in Chapter 3 a section that discusses being an active participant in your flying career and discovering why and how something is done. You have learned why, now you will truly understand how. Wait until a student asks you to explain why the nose drops in a turn or why they can't seem to track a rectangular course very well. Or maybe you've watched other pilots practice landings and realize that they don't feel right from the passenger seat. Perhaps the round out and flare are sloppy. Can you show, as well as tell that pilot how to improve?

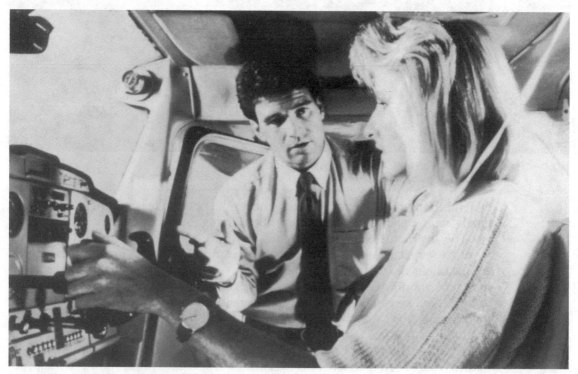

Fig. 6-5. A certified flight instructor (CFI) must know how to explain flying in a manner simple enough for any student to understand.

If you don't understand the innermost workings of every single maneuver necessary for a certificate, you'll never be able to teach them. Remember too, you'll need patience to watch a student make the same mistake over and over, hoping that the 14th time the correct action will sink in.

An instructor is a teacher of flying. And along with that task comes responsibilities. Humans grow up much like our parents, whether we would like to believe it or not. Our parents gave us a sense of right and wrong as well as teaching some of the finer things in life.

A flight instructor is the "parent" of a new student. You may be the only example of flying the student has contact with before obtaining a license. You know much more than students, and they stand in awe of you, even if they don't say it out loud. Realize this and work with it.

Think back to the number of teachers you've had during your public schooling and aviation education. What did you like about them? What did you dislike? Students will try to emulate their instructor. If your first few flight instructors were screamers, you will most likely try to handle your students the same way and that can be a loss for everyone. There are still plenty of old military instructors around who were trained by teachers who yelled at them until they performed correctly. Some instructors today still believe that if *they* could learn that way, so can others. Don't fall into that trap.

Was your instructor one of those that would look at you when you approached and say, "What did we do last time?" Think about how you felt. Perhaps that made you feel as if you were not important enough for your instructor to write anything down about, much less remember? Watch for the traps.

Some flight instructors teach because they enjoy—truly enjoy—flying and want to pass on what they've learned to others. Other flight instructors teach for personal gain, be it financial or flight experience. No matter what your reason for picking up a flight instructor rating remember you have a responsibility to the students. Someday they will fly away with what you've taught them filed deeply away in their brains. Make sure you've given them their money's worth.

CERTIFICATE AND RATING UTILIZATION

Some people believe there is a very strict barrier that extends between something they do for fun and something they might do to pick up extra money or earn a living. Some people will tell you that you'll destroy your love for flying if you try to make money. There may be some truth in that, but there is certainly an exception to all rules. I make money at flying and I still it as much as I ever did. And I'm not alone.

What could you do to earn money flying and what would you need? Certainly the first prerequisite to any flying job that will pay you for your efforts will be the commercial license with an instrument rating. With this and a few hundred hours flying time, you could tow banners or tow gliders. Each task is relatively simple, but involves enough challenge to be worthwhile. Banner and glider towing are not usually full-time endeavors for the business owner, although the demand for glider tow pilots is certainly greater than for banners.

Often you might be able to work part-time on weekends or during the week after your regular job. The banner towing company I worked for also had an electronic sign hung on the bottom of the plane: a billboard. We'd prepare the message and fly over a particular area, back and forth, with the sign flashing the message underneath. It was great fun and it kept the Citabria busy, sometimes 8 or 10 hours a day. Banners in the day, electronic signs at night. It all helped build my logbook hours so I could think of the next step up the ladder.

The list doesn't end here, though. Consider these other part-time or full-time flying jobs: sightseeing, crop dusting, powerline or pipeline patrol, or aerial photography. All of these jobs will require extra training and the company that hires you might provide the training. Crop dusting is almost a small rating course in itself because of the many techniques that are necessary to be a good duster pilot.

When you think you'd like to try one of these, drive out to the airports in your area and ask the managers of the various FBOs if they need anyone for these jobs or if they know anyone that does. Classified ads in a newspaper are one source for various jobs. Search through the yellow pages and if you want more sources of job information you can try Aviation Employment Monthly (800-543-5201) or Trade-A-Plane (615-484-5137). Contact FAPA at 800-JET-JOBS for their employment information.

You might get hired as a part-time copilot for a company twin-engine airplane; in this case the multiengine rating would come in handy. You might be able to log the multiengine time to help build experience, not to mention the practical experience you'll gain. The jobs are there if you look for them. I wandered around the airport for weeks, at all different hours, meeting people and asking questions. I met someone who heard of a company thinking about putting on a copilot in a Piper Seneca. The company wasn't sure it wanted one yet, but I found the owner of the firm and convinced him that if they did, I was their man. I eventually got a part-time job out of the deal and ended up flying the Seneca alone and picking up nearly 400 hours of multiengine experience in the deal.

Consider the possibility of being hired part-time as a charter pilot. This would pretty much depend upon the total number of hours of flying experience you have. The chances of being picked up to fly right seat on a light or medium twin when you have minimal experience is slim. You'll need more total time because, as a charter pilot, you're operating under a stringent set of flight rules, FAR Part 135. A pilot must be able to handle the airplane in instrument and emergency conditions.

Consider a part-time or full-time position as a flight instructor. This certificate will allow you to build time and experience and become involved with plenty of people from the aviation community. Keep in mind that you owe your students something for their money: a good hour of instruction. Make certain they get that before you get your head too far up in the clouds searching for the next rung on the ladder. Fly as a basic airplane instructor and perhaps pick up the instrument and multiengine instructor ratings as your experience builds. If you are working part-time at an FBO as a multiengine instructor, you'll be surprised at how quickly you'll be asked to go along on charter trips when they need an extra copilot. This is great to build multiengine time.

Perhaps you are thinking about more than flying little airplanes. Perhaps you seek the left seat of a Boeing 747. Guess what? Most of the airline pilots in the world started flying small airplanes and watching the aviation news for openings to help them move up the ladder. In the Midwest, a regional airline that provides feeder service to one of the major airlines is hiring copilots with 1000 hours total time and a commercial and instrument rating (Fig. 6-6). Proper experience will

Fig. 6-6. You could build multiengine time as copilot in a regional airliner like the Beech 1900.

boost you to a captain's position on a 19-seat airliner. This particular company also has provisions for the captains on the smaller aircraft to get priority on interviews for first officer (copilot) jobs on their DC-9s and Boeing 737s.

Today, more and more of the flying jobs are appearing in the regional side of the airline industry. The regional industry began flying very small, usually cramped twin engine turboprop aircraft twenty years ago, but that has all changed. Today's regional aircraft are much larger, much more sophisticated aircraft capable of passenger loads of 60 to 70 people. The cockpits of these aircraft, too, are loaded with all the state of the art electronics and flight management systems of larger jet airliners. A few of the regional airlines today have begun using smaller jet aircraft in regular passenger service. The pay is not quite as good as the major airlines, but this could be the way to get your foot in the door of an airline. Many of the regional airlines are owned by major airlines too, such as American Airlines' ownership of the American Eagle regional system.

Many of today's airline pilots were not trained in civilian life, however. Many began their flying careers in the air force, army, navy, marines or air national guard. Each offers terrific training programs, but the requirements are tough. You usually must be under 27 years of age when you sign on and must have a college degree, which most airlines also require. Remember, in the military, you won't be just a pilot if you make it through the training, you'll be an officer, a position that carries plenty of responsibility.

If you want to get a feel for what it's like to struggle through Officer Candidate School (OCS) with the navy, see the movie *An Officer and A Gentleman*. I'm told that's as real as it gets. If you make it through OCS you could be eligible for flight training. To see what navy pilots do, watch *Top Gun* starring Tom Cruise. It's a terrific action film and navy pilots tell me it's true to life.

If you really think a flying career in the military might be for you, contact your local recruiter. The training in the military is top-notch, but of course it does carry a pretty steep price for your time. If you were to leave after your hitch, you'd be a terrific candidate for most of the major airlines or corporate flight departments.

So what's it going to be? . . . a private license and flying up and down the coastline the rest of your life or a part-time job to start with and a career in aviation? Think about that the next time you see a Boeing 737 or corporate Gulfstream IV or even an F-16 fly over your house. Where would you rather be?

7

Associations, magazines, & the personal computer

RECALL THAT CHAPTER 2 MENTIONS some of the best times could be found around other people who enjoy aviation. Another way to enjoy aviation could be through some of the major aviation organizations, as well as some of the aviation publications available to you in a number of different ways.

THE NINETY-NINES

The Ninety-Nines is an international organization of licensed women pilots around the world, although the vast majority of the 6800 members live in the United States. The Ninety-Nines formed in 1929 at Curtis Field, Long Island, New York, with noted aviatrix Amelia Earhart elected as the organization's first president. The stated purpose of the Ninety-Nines is "to engage in strictly educational, charitable, and/or scientific activities and purposes, and particularly to promote aeronautical science by such means as is not inconsistent with the educational, charitable, and scientific purpose of the corporation."

More than 300 educational programs in 1993 included flight instructor revalidation courses, fear of flying clinics for shaky airline passengers, airport tours for school children, aviation education workshops for school teachers, talks to service clubs, and copilot courses for apprehensive light-plane passengers.

Much volunteer work is performed by the Ninety-Nines' membership and their aircraft for the Red Cross and for cancer victims through the Corporate Angel Network (CAN) each year. Members often serve on airport commissions and give introductory airplane rides to scout troops. Over the years the Ninety-Nines have given much back to the country and the aviation community. If you are a woman pilot, or a woman student pilot (who are called Sixty-Sixes), you should investigate membership in the Ninety-Nines. The address is:

Ninety-Nines, Inc.
International Headquarters
P.O. Box 59965
Will Rogers Airport
Oklahoma City, OK 73159

Fig. 7-1. Vintage photo of two Ninety-Nines members.

The 99s, Inc., International Women Pilots, Archives, Oklahoma City, OK

AIRCRAFT OWNERS
AND PILOTS ASSOCIATION

The Aircraft Owners and Pilots Association (AOPA) will celebrate its 55th anniversary in 1994. AOPA is often referred to as the "Voice of General Aviation" and currently represents members who own or fly general aviation aircraft. Since its inception, AOPA has fought hard for the rights of pilots to fly, something that the

Fig. 7-2. The AOPA Fly-In.

government and the airlines often find to be in conflict with what they see as the future of aviation. The airlines to this day will often view general aviation flying as just a bunch of people flying little airplanes. General aviation pilots are often seen as the obvious cause of accidents and are automatically assumed by the press to be flown by inexperienced pilots. On more than a few occasions, AOPA has lobbied in Congress and through the news to change those ideas. AOPA President Phil Boyer said:

> Since its earliest days, flying has inspired fellowship among those who take to the skies. That spontaneous feeling linking pilots as comrades is reinforced today by practical reasons of mutual help.
>
> Through the Aircraft Owners and Pilots Association, AOPA, pilots have banded together in the cause of general aviation for over half a century. Today we represent more than 300,000 members, some 60% of all active pilots in the United States and our membership continues to grow.
>
> With this unified strength, AOPA is general aviation's principal watchdog, defender and voice of progress. Our government affairs team is the strongest aviation advocate in Washington. Our technical affairs specialists constantly

monitor technical developments and FAA regulations in the interests of the individual pilot. Our regional representatives provide AOPA expertise and leadership at the local level throughout the country.

Each member is a part of this force for general aviation. Besides helping to keep flying free of unreasonable restriction and cost, each member receives the leading monthly magazine in the general aviation field, *AOPA Pilot*, and our annual directory and source book, AOPA's *Aviation USA*. Members also benefit from special rates and discounts on personal and aircraft insurance, travel, credit card accounts and purchases of aviation equipment and supplies.

In this book, Rob Mark has done a fine job of expressing "the joy of flying." New pilots who join AOPA will experience both the joy of flying and the satisfaction of participating with fellow pilots in preserving this priceless national resource, general aviation.

Thanks for the kind words, Phil.

AOPA's Air Safety Foundation continues to provide pilots with training programs to keep them abreast of what is happening in the aviation world. The Air Safety Foundation also teaches specialized courses, such as mountain flying and the famous "pinch-hitter" course that teaches the spouse of a pilot enough about flying an airplane to take over and land in an emergency. AOPA offers an aviation chart service for members to make certain they are flying with the most up-to-date information possible. Contact:

Aircraft Owners and Pilots Association
421 Aviation Way
Frederick, MD 21701

EXPERIMENTAL
AIRCRAFT ASSOCIATION

From the time I first heard of the Experimental Aircraft Association (EAA), I knew there was something big going on. I was about 15 at the time and was lucky enough to get a volunteer job at EAA's Rockford, Illinois, air show helping park airplanes. I spent an entire week, wandering around the aircraft parking area. They gave me big orange paddles to direct airplanes the likes of which I'd never seen before, even though at that point I'd read just about every book on airplanes ever written. My reward that summer of 1964 was a ride in the back seat of a P-51 Mustang by some great guy whose name I no longer remember. I was hooked.

For those of you who are new to general aviation, the EAA no longer holds its air shows at Rockford. The field was just too small to accommodate the thousands of airplanes, pilots, and families who attend each year. The association now holds the air show at Oshkosh, Wisconsin. It is the largest air show and aviation display of its kind in the world. In fact, during the "Week of Oshkosh," as it has become known, the Wittman field becomes the busiest airport in the world—even busier than Chicago O'Hare. The air show reached a milestone in 1988. EAA had to deny aircraft permission to land and park on the field because there simply was not a single bare patch of grass left to park another airplane.

EAA is the largest organization of homemade airplane aficionados in the world. A pilot certificate or aircraft ownership is not required for membership. EAA caters to everyone who loves aviation. The basic goal of EAA is to support sport aviation. Thousands of EAA members worldwide have built their own aircraft since the organization was formed some 35 years ago.

EAA maintains a strong voice in Washington to help the cause of amateur builders of aircraft as well as to help those who own aircraft produced by manufacturers. For anyone interested in building an aircraft or merely the camaraderie of those who do, the EAA maintains more than 700 local chapters around the world (Fig. 7-3). There is bound to be one near you. EAA has something for everyone in one of its divisions. In EAA's own words, "The Antique/Classic Division helps EAA members rediscover the epic of flight through the restoration and preservation of the great flying machines of yesteryear."

EAA's International Aerobatic Club (IAC) sponsors worldwide competition in various categories of aerobatics. Many are open to beginners because the division believes "aerobatics are for everyone. Relish the unmatched thrill of progressing from simple loops and rolls to intricate multiple inside/outside maneuvers." There can be no doubt that aerobatic knowledge makes a better pilot. Within just a

Fig. 7-3. EAA chapter meeting.

EAA/Jim Koepnick

few lessons you'll find yourself feeling comfortable in attitudes that you earlier might have been afraid of. Knowledge is security to you as a pilot; aerobatic knowledge is invaluable.

EAA's Warbirds Division includes pilots and enthusiasts who are devoted to preserving World War II aircraft. Each year at Oshkosh, one entire portion of the airport is devoted to warbirds. Oshkosh 1988 saw a record number of T-6s, nearly 100 of them, arrive at Oshkosh at the same time. Other individually-owned aircraft in the group include numerous P-51s, F-4U Corsairs, P-38 Lightnings, and T-28s. For the person who is a real WWII airplane lover, there is hardly a better place to be than in a local Warbirds chapter.

EAA's Air Adventure Museum near Wittman Field was founded entirely by private funding and operates almost 365 days a year. The museum celebrates the history of experimental aircraft, from the beautiful reproduction of the Wright Brothers' "Flyer," to a replica of Charles Lindbergh's "Spirit of St. Louis." The Spirit replica has been flown on a cross-country tour around the United States promoting general aviation. The museum includes many of the early model warbirds, such as the XP-51 that led to production of the P-51 Mustangs of WWII. A museum tour will show you some of the most famous experimental aircraft designed and built by some of the famous EAA homebuilders, such as Paul Poberezny's Baby Ace and Ray Stits' Sky Baby, reportedly the smallest flying manned aircraft in the world.

In 1991, the EAA Foundation conducted a survey of long-time members of EAA to help determine priorities for the organization's future. An overwhelming majority (92 percent) said the organization's primary objective should be to involve more young people in aviation. After several months of coordination by members of EAA and Foundation Boards of Directors, EAA management, staff and volunteers, the Young Eagles Program was formally announced at a news conference in Washington, D.C., on May 13, 1992 and officially launched during the 1992 EAA Fly-In Convention in Oshkosh, Wisconsin. Academy Award winning actor and pilot Cliff Robertson serves as the program's first Honorary Chairperson.

The mission of the Young Eagles program is to provide a meaningful flight experience for 1 million young people (primarily between the ages of 8 and 17) by the year 2003 — the 100th anniversary of the Wright brother's first powered flight and the 50th anniversary of the Experimental Aircraft Association.

To attain its goal, the Young Eagles will include the following activities:

EAA volunteer pilots will identify young people who wish to participate in the Young Eagles Program and take them for a demonstration airplane ride.

Throughout the next decade, Young Eagles Flight Rallies will be held at designated airports across the country. During these events, parents can bring their children to experience the exhilaration of flight. Contact:

Ed Lachendro, Young Eagles
Coordinator at EAA
P.O. Box 3086
Oshkosh, WI 54903
800-322-2412

SOARING SOCIETY OF AMERICA

The Soaring Society of America (SSA) is a nonprofit association of 16,000 soaring enthusiasts from Hawaii to Florida that was founded in 1932 by a small group of pilots to organize a national soaring (glider) championship.

SSA has grown, becoming involved in all aspects of the sport, from initial training to international competition. It is a division of the National Aeronautic Association, and SSA oversees state and national soaring records and awards soaring proficiency badges.

The Soaring Society of America's monthly journal, *Soaring*, has up-to-date news on sailplane developments and soaring activities along with articles of interest to novice and expert pilots. Soaring Society of America workshops, seminars, and meetings address the special needs of glider pilots. Twenty-six volunteer directors and 55 volunteer state governors represent members on vital issues like airspace restrictions, insurance costs, pilot certification, and safety.

Membership includes the monthly journal *Soaring* as well as a membership handbook. A directory of soaring sites and organizations around the U.S. is available for $3. Contact:

The Soaring Society of America, Inc.
P.O. Box E
Hobbs, NM 88241-1308

THE CIVIL AIR PATROL

The Civil Air Patrol (CAP) is a civilian auxiliary of the United States Air Force. CAP members are not members of the United States Air Force, but are entitled to wear the air force uniform under certain conditions. The CAP mission has two levels of membership: senior, or cadet (under 18). Three CAP missions are search and rescue for aircraft, boats, and other missing persons, emergency services, and aerospace education.

Whenever an aircraft is lost or reported overdue through either a flight plan notification or other method, the search and rescue forces of the country are notified through the U.S. Air Force Search and Rescue Headquarters at Scott Air Force Base in Illinois. The duty officer at Scott will determine which wing of the CAP (each state is a wing) will handle the call. Wing officials contact one or more local squadrons that move into the field on foot and in the air to conduct the search and rescue operations. Only mission-qualified pilot members of CAP are allowed to participate in these operations.

Another important phase of CAP is emergency services. In times of natural disaster, CAP works closely with Civil Defense to evacuate the injured or speed medical supplies to normally inaccessible areas. Cadets and senior members will often provide additional assistance in the form of communications or just plain manual labor to help those in need. When on an authorized mission for disaster relief or search and rescue, CAP members are reimbursed for costs, such as the fuel and oil necessary to fly their aircraft.

CAP's aerospace education effort maintains interest in aviation and space for people under age 18 (Fig. 7-4). Seminars cover model airplane and rocket building

Fig. 7-4. CAP cadet in a jet simulator.

through flight training in gliders during the CAP summer flight programs. Camp outs, called bivouacs, specialize in navigation and search and rescue training.

CAP is an auxiliary of the U.S. Air Force and wearing an air force uniform is optional in certain CAP situations and required in other situations. If you become a senior member (over 18) you may wear the uniform if you meet certain USAF standards and may progress through the various military-style CAP ranks, beginning with 2nd lieutenant.

Cadets begin with the rank of airman basic and move through various positions with more and more responsibility. They could eventually reach the cadet officer ranks and possibly become the cadet commander.

If you enjoy flying and working with young people in a military environment, the volunteer Civil Air Patrol is certainly worth checking out. Check at your local airport to see if a squadron exists or write to:

Civil Air Patrol
National Headquarters
Building 403
Gunter AFB, AL 36414

AVIATION/SPACE WRITERS ASSOCIATION

If you enjoy aviation and have a knack for writing about it, in any form, the Aviation/Space Writers Association (AWA) is for you. Numbering about 600 members, AWA has members in every portion of the nation's media, whether it be television, radio, or print. AWA is broken into regions and often into chapters, depending upon local member interest.

The thrust of AWA is to promote the accurate reporting of aviation- and space-related stories and to help members of the press to accurately report on aviation and space topics. AWA holds an annual "news conference" in a different city across the country. The conferences usually include visits from top civilian and government aviation officials, as well as panels to discuss the problems that aviation faces on a day-to-day basis.

If you're a journalist or plan to become one, consider the Aviation/Space Writers Association. Special memberships for student journalists are also available. Write:

Aviation/Space Writers Assoc.
Suite 1200
17 S. High Street
Columbus, OH 43215
614-221-1900

CONFEDERATE AIR FORCE

Contrary to what the name might indicate, the Confederate Air Force (CAF) is not an underground defense league. CAF is a group of people who always wanted to be around WWII airplanes and feel strongly about preserving this particular part of America's history. The Confederate Air Force states it goals more precisely:

1. To preserve, in flying condition, a complete collection of combat aircraft [that] were flown by all military services of the United States in World War II.
2. To provide museum buildings for permanent protection and display of these aircraft. To establish the "Combat Airman Hall of Fame" as a tribute to the thousands of men and women who built, serviced, and flew them.
3. To perpetuate in the memory and hearts of all Americans the spirit in which these great planes were flown for the defense of our nation.
4. To establish an organization having the dedication, enthusiasm and Esprit de Corps necessary to operate, maintain and preserve these aircraft as symbols of our American Military Aviation Heritage.

CAF is an all-volunteer organization, but all work on CAF aircraft is performed by members under the direct supervision of qualified mechanics.

Two years after formally organizing in 1961, the Confederate Air Force completed its collection of at least one of each American fighter used during WWII. The bomber squadron was completed in 1972.

The fighter squadron includes an F6F Hellcat, F8F Bearcat, F4F Wildcat, F4U Corsair, SBD Dauntless, SB2C Helldiver, TBF Avenger, P-40 Warhawk, P-38 Lightning, P-39 Aircobra, P-63 Kingcobra, P-51 Mustang, and P-47 Thunderbolt, and P-82 Twin Mustang (Fig. 7-5).

Confederate Air Force

Fig. 7-5. Confederate Air Force Curtis P-40 and Grumman F4 Wildcat.

The bomber squadron includes a PBY Catalina, B-29 Superfortress, B-17 Flying Fortress, A-20 Havoc, A-26 Invader, B-24 Liberator, B-26 Marauder, B-25 Mitchell, and a Lockheed Hudson. Over the years, the CAF has also acquired transports like the C-46 Commando and the C-47 Skytrain. They've also located a British Supermarine Spitfire, a German He.111 bomber, a Junkers Ju.52 Transport, and a Messerschmitt Bf.109. The remainder of the "Ghost Squadron," as the CAF calls the fleet, is military trainers like the early Stearmans and T-6s.

The CAF maintains a museum at Midland International Airport, Texas, where 137 aircraft representing 60 different types are housed. Along with the museum is an extensive collection of aircraft memorabilia including more than 50,000 artifacts. The CAF attends numerous air shows each summer in various parts of the country where the populace can see some of the CAF-restored aircraft close up. Anyone may join the CAF and receive the CAF *Dispatch* magazine. Members may also participate in the activities of the local CAF chapter. Write:

Confederate Air Force
P.O. Box 62000
Midland, TX 79711-2000
915-563-1000

BALLOON FEDERATION OF AMERICA

A balloon was the first vehicle to carry a man aloft, which occurred in late eighteenth-century France. The Balloon Federation of America (BFA) carries on the oldest form of flying. Members "dedicate themselves to the safety, enjoyment and advancement of the sport of ballooning, as well as the promotion of friendship among balloonists around the world."

BFA has taken an active role in the education of its pilot and flight crew members through local and national safety seminars regarding fire safety, weather, landowner relations, crew techniques, and fuel management. The organization also maintains an extensive library of written, video, and film programs for its members. BFA encourages advancement of flying skills through the Pilot Achievement Award Program. The organization also maintains international ties in the ballooning community through the National Aeronautic Association and the Federation Aeronautic Internationale. A monthly in-house newsletter supplements the full-color quarterly magazine *Ballooning*. The organization supports the National Balloon Museum of lighter-than-air artifacts in Indianola, Iowa. Write:

Balloon Federation of America
P.O. Box 400
Indianola, IA 50125

UNITED STATES PILOTS ASSOCIATION

The focus of the USPA is to foster the development of strong and effective statewide pilots' associations, each of which supports the development of forceful local chapters to serve the aviation interests of their members and their community. USPA encourages mutual support between the states, aviation safety and pilot education. Each affiliated state organization and the individual members have representation on the board of directors, thus making USPA genuinely a member controlled organization. Sixteen state pilot associations currently belong to USPA. Contact:

USPA
483 S. Kirkwood Road, Suite 10
St. Louis, MO 63122

PILOTS INTERNATIONAL ASSOCIATION

The Pilots International Association has been around since 1965 providing the general aviation pilot with a wide range of services and benefits all designed to make flying safer and more fun. Some of the services include PIA's quarterly newsletter, *Flight Line*, which is filled with membership news, safety updates, a regular article on FARs as well as benefit information. Your membership also includes a subscription to *Plane & Pilot*. The association also offers a free job listing service if you are looking for work. The want ad will be read by thousands of members, so this alone could make your membership worthwhile. There is also a life insurance program as well as special discounts on flying items such as computers and books. Finally, PIA offers its members free audio-visual tape rentals on dozens of aviation subjects. Contact:

PIA
4000 Olson Memorial Hwy.
Minneapolis, MN 55422
612-588-5175

UNITED STATES
ULTRALIGHT ASSOCIATION

We haven't spoken of ultralights before. Ultralights are flying machines that do not require either the aircraft or the pilot to be licensed but are actually capable of flying. But before you go off deciding this is for you, let me tell you about a few of the restrictions, and there are many. The empty weight of the craft may not exceed about 250 lbs nor can it carry more than 5.5 gallons of fuel. You cannot fly the craft into controlled airspace nor near a controlled airport, over congested airspace, at night or whenever the visibility is less than three miles. What's left is absolutely fair weather flying. But if you are trying to reach for the sky on just a few dollars, call the United States Ultralight Association for details about this sport. They'll tell you about where to find and fly ultralights, how much they cost, what regulations you must observe and even give you a subscription to their nifty magazine, *Ultralight Flying*. Contact:

U.S. Ultralight Assoc.
P.O. Box 667
Frederick, MD 21705
301-695-9100

ANTIQUE AIRPLANE ASSOCIATION

If you've ever wondered what happened to airplanes of the biplane era, you'll be happy to know they return to Bartlesville, Oklahoma, each summer as part of the Antique Airplane Association Fly-In. AAA was the first organization dedicated to keeping old airplanes alive, formed in 1953. AAA has 42 chapters, and eight affiliated clubs for specific airplanes, around the country.

The association owns "Antique Airfield," at Blakesburg, Iowa. It's often used for antique and classic aircraft get-togethers during the year, such as the annual Stearman Fly-In. The association staff is also around to assist rebuilders with help in making a flying machine out of a bunch of spare parts.

AAA publishes a quarterly newsletter. Write:

Antique Airplane Association Inc.
Antique Airfield
Route 2, Box 172
Ottumwa, IA 52501

SEAPLANE PILOTS ASSOCIATION

If you've ever thought about owning or renting a seaplane and flying to a lake cabin in Wisconsin, for a weekend of fishing, or simply goofing off, the Seaplane Pilots Association is for you.

Most airplanes are designed to land on a hard surface or grass. Seaplane pilots forget those boundaries (Fig. 7-6). Many of their aircraft are amphibious and can use either a standard runway or the lake. The Seaplane Pilots Association (SPA), which is affiliated with AOPA, was formed in 1972 to protect the water-landing rights of all seaplane pilots, by removing restrictions and opening new waters.

Fig. 7-6. Amphibious seaplanes open up many new landing sites.

SPA helps members cope with seaplane ownership, maintenance, pilot certification, flight training, and legal referrals. A quarterly magazine, *Water Flying*, and a comprehensive annual publication of the same name are published by SPA. Additionally, the association offers members a *Seaplane Landing Directory*, a guide to landing sites around the nation as well as any restrictions that might apply in those states. The association also sponsors 12 Fly-Ins a year at different locations. Write:

Seaplane Pilots Association
421 Aviation Way
Frederick, MD 21701.

NATIONAL ASSOCIATION OF FLIGHT INSTRUCTORS

Flight instructors might find the National Association of Flight Instructors (NAFI) worthwhile. NAFI offers instructor refresher clinics and publishes a newsletter. NAFI was also instrumental in establishing professional liability insurance to cover instructors. NAFI also operates a job-placement service. Write:

NAFI
Ohio State University
P.O. Box 793
Dublin, OH 43017

THE WHIRLY-GIRLS, INC.

The Whirly-Girls was organized in 1955 with 13 women helicopter pilots in the United States, France, and West Germany to support women helicopter pilots. Today the Whirly-Girls number 650 in 22 countries.

The organization offers significant scholarship programs, like the $4000 Doris Mullen Memorial designed to assist a deserving female commercial pilot with the funds needed to upgrade to a helicopter rating. The organization also recently added another $4000 scholarship in the name of Sheila Scott, who died in October 1988.

Whirly-Girls has an ongoing program to help develop more hospital heliports around the country. Write:

The Whirly-Girls, Inc.
1619 Duke Street
Alexandria, VA 22314-3406

NATIONAL AIR AND SPACE MUSEUM

Certainly one of the premier historical aviation locations in the nation is the National Air and Space Museum, a part of the Smithsonian Institution in Washington, D.C. You can easily spend a few days there viewing everything from the Wright brothers' Flyer to modern day military and civil aviation, plus displays on the space effort since the days before Sputnik. Charles Lindberg's original *Spirit of St. Louis* resides here too. The Air and Space Museum offers a membership with benefits similar to those offered in the standard Smithsonian membership. Write:

National Air and Space Museum
Smithsonian Institution
Washington, DC 20560

AMERICAN AVIATION
HISTORICAL SOCIETY

The American Aviation Historical Society strives to share information based on members' research. A quarterly newsletter features historical information on practically every area of aviation, as well as reviews of aviation books and videos. The society also maintains a large library of rare negatives and slides. Write:

American Aviation
Historical Society
2333 Otis Street
Santa Ana, CA 92704

WORLD WAR I AEROPLANES

Another organization is WWI Aeroplanes for "collectors, restorers, replica builders, historians, and modelers all over the world. . . ." The organization publishes two different magazines: *WWI Aero* specifically highlights aircraft from the period 1900 to 1919; *Skyways* covers the years 1920 to 1940. Each issue is crammed with technical drawings and data, as well as photographs and news on current projects in the works amongst readers. Write:

World War I Aeroplanes, Inc.
15 Crescent Road
Poughkeepsie, NY 12601

LAWYER–PILOTS BAR ASSOCIATION

This group includes members who are primarily interested in aviation law, lawyers who have an interest in aviation safety and the largest group, those who are lawyers and pilots and enjoy professional fellowship with like-minded individuals. There is a special category for law students and a category for patrons that contribute to the financial support of the association's efforts to enhance aviation safety. A subscription to the association's publication, *LPBA Journal*, is included with membership. Contact:

LPBA
500 E. Street, S.W.
Suite 930
Washington, D.C. 20024
202-863-1000.

"AIRPLANE-TYPE" ORGANIZATIONS

If you own an airplane, there is probably a club to support your airplane (Fig. 7-7). At the very least, someone publishes a newsletter to support specific aircraft.

The Cessna Owner Organization publishes quite a fine magazine designed to help owners find new and effective ways to maintain their Cessna aircraft. A recent issue included a pilot report on an early model 310. The magazine is also chock full of sources for parts and services for the Cessna owner. This organization also publishes *Pipers* magazine at the same address. Call 800-331-0038.

Cessna Owner Organization
121 N. Main Street
P.O. Box 337
Iola, WI 54945

Piper Owner Society
P.O. Box 337
Iola, WI 54945
800-331-0038

American Bonanza Society
Mid-Continent Airport
P.O. Box 12888
Wichita, KS 67277

Mooney Aircraft
Pilots Association
314 Stardust Drive
San Antonio, TX 78228

Fig. 7-7. A typical "pride-and-joy" of the American Bonanza Society.

American Navion Society
P.O. Box 1175
Municipal Airport
Banning, CA 92220-0911

Cherokee Pilots' Association
P.O. Box 716
Safety Harbor, FL 34695

International Cessna
120/140 Association
P.O. Box 830092
Richardson, TX 75083-0092

Cub Club
P.O. Box 2002
Mt. Pleasant, MI 48804-2002

Luscombe Association
6438 W. Millbrook
Remus, MI 49340

Replica Fighters Association
2789 Mohawk
Rochester, MI 48064

Taylorcraft Owners Club
12809 Greenbower Road
Alliance, OH 44601

National Aeronca Association
266 Lamp and Lantern Village
Chesterfield (St. Louis), MO 63017

American Yankee Association
3232 Western Drive
Cameron Park, CA 95682

David E. Neumeister
Aircraft Newsletters
5630 South Washington
Lansing, MI 48911-4999

David E. Neumeister Aircraft Newsletters cover just about every other make of aircraft that isn't in the list above. The newsletters report on many aircraft: Cessna 310/320, 336/337, Varga, Beech Baron, Musketeer, and Skipper, Helio, Maule, Rallye, Piper Tomahawk, Arrow, Malibu, and Cherokee Six.

THE PILOT & THE PC

Today, no pilot can be considered informed if he or she is unaware of some of the sources of information and services available to them through the use of a personal computer (PC). A PC connected through a modem in your unit can put you in touch with thousands of people in the outside world via the online databases who are plane nuts too, not to mention many who are professional pilots. The selection of equipment is beyond our scope, but once you have one, you'll find yourself able to electronically keep your logbook, or actually run and log simulator time from a program run on a PC or even study for your next FAA written exam through question and answer sessions that take you through the material much faster as well as making it more interesting. Let's take a look at some of the items you can find.

First are *CompuServe* and *AOPA ONLINE*. Both are called on-line databases, CompuServe being the largest with just over 1 million members worldwide. *AOPA ONLINE* is significantly less but began service only in early 1993. Both services are accessed through a local phone call from your PC. *CompuServe* connects to a database with hundreds of different subjects, forums, and software available for download (Fig. 7-8). The most popular stop for pilots on *CompuServe* is the AVSIG, or aviation special interest group. Here a pilot can learn about local weather as well as talk to others about navigation or safety questions. Other subjects include air traffic control, places to fly, aviation computer programs, training and careers, want ads and much more.

AOPA ONLINE, although considerably smaller, is devoted entirely to aviation, 100 percent through the people who run *AOPA ONLINE*, the Aircraft Owners & Pilots Association. *AOPA ONLINE* also contains a number of forums as well as a method to post specific questions for any and all to see and answer, like "Anyone know where the best place is to find a good Cessna 120?" Within days, often hours depending upon the question, other on-line pilots will send along bits and pieces of information to help you in your quest. If you haven't played with a PC and on-line services, head to your computer store for a demonstration. Both *CompuServe* and *AOPA ONLINE* offer their own software to make navigating these databases a bit easier.

When I started flying 26 years ago, I don't think I ever realized just what a pain in the neck keeping up a logbook would be. I was making entries every day or so and constantly adding numbers incorrectly, thereby making a mess out of my logbook or

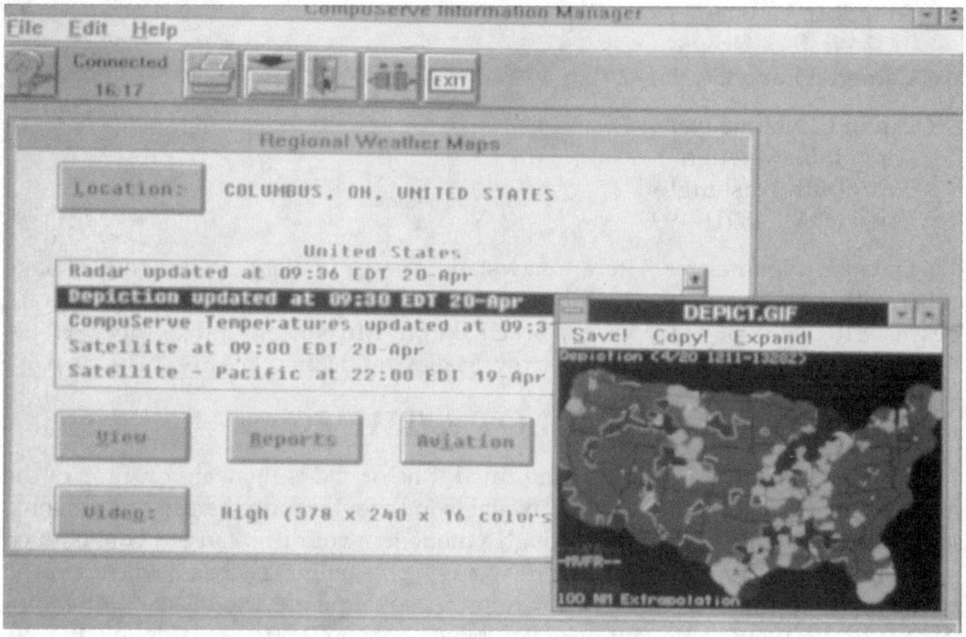

Fig. 7-8. CompuServe allows you to display current weather maps.

the whiteout factory as I sometimes called it. But then one day, I was breezing through the software offered on *CompuServe* in AVSIG and I saw a demonstration program available called *Aerolog*. The writer had saved a simplified version of his program on CompuServe and anyone could download it for free to try it out. I did.

Within a few hours of typing in the entries from my first logbook (I had three), I saw the simplicity of it all. This computer program remembered all the tail numbers of all the aircraft I'd ever flown. When I typed N9MK the second time and the 102nd time, the computer did the work and put the aircraft make and model into the log entry. It added all the columns instantly. It told me in a fraction of a second how much time I had already logged in a Cessna 310 or a Piper Lance. Writing in all the numbers on to the FAA recommendation sheet for a checkride is usually a nightmare because of the way the time must be broken down. *Aerolog* makes it a walk through the tulips. Certainly there are other logbook programs, but I wouldn't give up my *Aerolog* computerized logbook for anything (Fig. 7-9). Call *Aerolog* at 800-336-1204.

How about if you decide you'd like to practice some approaches since you're about to pick up your instrument rating. Or maybe you already own an instrument rating and need a way to keep current, normally an expensive proposition. With some of the new generation flight simulator programs that you install on your PC, you'll find yourself able to simulate flying aircraft from a Cessna 172, to a WW II P-51, to a Learjet to a Concorde. Two versions of simulator programs I've played with and enjoy perform similar missions, but to differing degrees of realism.

Each uses a simulated cockpit that appears on the CRT of your computer. Through the use of a device called a joystick that looks like the control stick to an

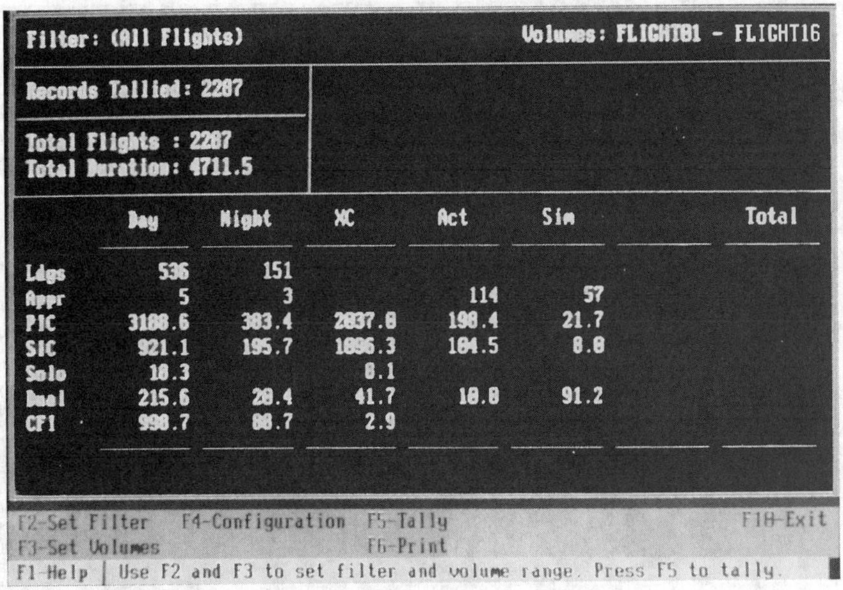

```
Filter: (All Flights)                          Volumes: FLIGHT01 - FLIGHT16

Records Tallied: 2287

Total Flights : 2287
Total Duration: 4711.5

              Day      Night       XC        Act       Sim         Total

Ldgs          536       151
Appr            5         3                   114        57
PIC        3188.6     383.4     2837.8      198.4      21.7
SIC         921.1     195.7     1896.3      184.5       8.8
Solo         18.3                   8.1
Dual        215.6      28.4       41.7       18.8      91.2
CFI         998.7      88.7        2.9


F2-Set Filter    F4-Configuration  F5-Tally                    F10-Exit
F3-Set Volumes                     F6-Print
F1-Help │ Use F2 and F3 to set filter and volume range. Press F5 to tally.
```

Fig. 7-9. AEROLOG keeps track of your logbook on your PC.

airplane and plugs right in to your PC, you can actually fly your aircraft. Stick left the craft banks left, stick back up comes the nose. Want to try a loop in a P-51, make sure you have enough airspeed first, lower the nose slightly as you pick up speed, haul back on the stick and watch the horizon out your windows turn into blue sky as you go over the top. Don't forget to pull the power back on the way down or you'll pull so many G's you could tear the wings off or even pass out (the screen starts to turn red).

Personally I own three of these crazy simulators. One, Chuck Yeager's Advanced Flight Trainer, allows you to fly not only the P-51 but also fly formation aerobatics and even includes dual flight training if you don't already know how to fly.

Microsoft's Flight Simulator has its benefits too, especially since it includes a scenario beginning at the approach end of runway 36 at Chicago Meigs where I've flown into and out of hundreds of times. I can take off, set up the radio properly and turn northwest and fly directly to and shoot the 27 right ILS instrument approach at O'Hare. Wait til you try landing one of these things. That can be the best fun.

Another program I've recently begun learning is Flight Deck Software's Instrument Flight Trainer. What attracted me to this system was the realism of the cockpit. I can set it up as an A36 Bonanza like I often fly with an HSI and an RMI (two fairly sophisticated pieces of navigation equipment). Besides being able to choose a number of different aircraft and regions of the country to fly in, IFT will allow for easy connection of control wheels in place of the computer joystick as well as a set of electrically controlled rudder pedals for even more realism.

I'd better stop here though because as these systems become more and more real, you may not want to go to the airport very often. But seriously, if you haven't

Fig. 7-10. You can keep IFR current on your lunch hour.

made the transition to a personal computer yet, try one. They're the best for an afternoon's fun when you can't really make it out to the airport (Fig. 7-10). Try calling for the *Flight Computing Catalog* at 800-992-7737 for a list of some of the best flight simulator and related programs and accessories.

MAGAZINES

Every group has its special list of publications, and fliers are no exception. Below you'll find a list of some of the top publications and where to write to subscribe or pick up a sample issue. The magazines that can be found on the newsstand are marked with an asterisk(*).

Air & Space
National Air and Space Museum
P-700
Washington, DC 20560

Air Line Pilot
535 Herndon Parkway
P.O. Box 1169
Herndon, VA 22070

AOPA Plot
421 Aviation Way
Frederick, MD 21701

*Air Classics
Challenge Publications, Inc.
7950 Deering Avenue
Canoga Park, CA 91324

*Air Progress
Challenge Publications
7950 Deering Avenue
Canoga Park, CA 91324

Aviation Week & Space Technology
1221 Avenue of the Americas
New York, NY 10020

FAA Aviation News
FAA Flight Operations
AFO 87-DOT
Washington, DC 20591

Flying
DCI Communications
1575 Broadway
New York, NY 10036

Plane & Pilot
Werner & Werner Corp.
16000 Ventura Boulevard
Suite 800
Encino, CA 91436-2782

Professional Pilot
3014 Colvin Street
Alexandria, VA 22314

Sport Aviation
EAA Wittman Field
Oshkosh, WI 54903-3086

Flight Training
405 Main Street
Parkville, MO 64152

Kitplanes
Fancy Publications, Inc.
P.O. Box 6050
Mission Viejo, CA 92690

Private Pilot
Fancy Publications
P.O. Box 6050
Mission Viejo, CA 92690

Rotor & Wing International
PJS Publications, Inc.
News Plaza
P.O. Box 1790
Peoria, IL 61656

Trade-A-Plane
410 W. 4th Street
Crossville, TN 38555

VIDEO MAGAZINES

A new era of video aviation magazines has recently begun in America. They offer the viewer the chance not just to read about aviation, but also to experience it visually. Some of the new videos offer practical tips on flying. Others take you right into the cockpit of a P-51 Mustang or an F-16 for a pilot's view of those machines. The possibilities in this medium are endless and they're just getting started.

Aviation Week Video Club
McGraw-Hill Aerospace & Defense Group
P.O. Box 308
Mt. Olive, NJ 07828

Aviation Week & Space Technology tapes look at different aspects of flying: commercial, military, and space. Current titles include "Space Shuttle: The Recovery," "Test Pilot," "Paris Air Show," and "Flight Deck."

These magazines and associations all give you, the general aviation pilot, the opportunity to participate in and learn more about the world of flying. Take advantage of them.

8

Time to move up?

PERHAPS BY NOW YOU'VE GAINED some VFR flying experience in a Cessna 150 or perhaps an instrument rating in a Piper Cherokee or Cessna 172. One of my instructors used to say the reason people own airplanes is because they want to go somewhere . . . fast! Perhaps you've realized your Cessna 150 or Cherokee is not the fastest machine in the world, nor can it carry a tremendous load over a long distance. As with automobiles, each airplane has a particular place in the market and cannot be all things to all people. If you want to fly at 200 mph, you're wasting your time in a Cherokee 140. If you want to carry a family of six plus baggage, you might as well give up on a Cessna 172 because it just isn't going to work.

As with automobiles, there is always something just a little bigger and a little faster, but the problem is whether or not you can afford it. How much are you willing to spend to pick up the extra speed?

THINK BEFORE YOU BUY

Certainly no better piece of advice was given a new pilot than to think before you buy and do not buy the first airplane you see. In aviation, a little knowledge can be dangerous. You'll find within a few months of picking up your license or some additional rating that you'll probably join one of the organizations listed in Chapter 7. The only problem with all the information you'll pick up in some of these organizations is you'll start thinking about LORANs, RNAVs, and HSIs before you completely understand what these items are or how they might enhance your flying abilities. It can be the same way with airplanes.

You're bound to meet other pilots who will have more sophisticated airplanes than you, and after a few rides in them you're going to want one. It's pretty tough to sit in the right seat and watch a pilot retract the landing gear or push the throttle in until this new machine accelerates to 185 mph. It doesn't take too much for you to see yourself getting to meetings faster than before. It also isn't going to take much to realize that a 10-hour trip to Disney World in your Cessna 150 could be cut to 5 hours in one of these fancy new machines.

How do you know if the old buggy you picked up your private and instrument rating in is no longer effective? Maybe it would be better to add some other, more sophisticated avionics to your current airplane than to go through the trouble of buying another airplane and learning new systems all over? This is a pretty tough decision to make on your own, but it certainly is not a decision a writer 5000 miles away can make for you. What I will do though, is give you a few tips about what to consider when the move-up bug bites.

The most important item on the list is whether you need this airplane or want it. If, of course, you have money to spare, this discussion is purely academic. But, if you fall into the range the other 95 percent of the country's pilots are in, money must be a consideration. Look at how many hours you've flown since you bought the first airplane. Divide up all the costs and figure out what it cost per hour. Realize that a more complex airplane, no matter what shape it's in, is going to cost more per flying hour.

If the machine has a variable-pitch propeller (called a constant-speed prop), this will add to the cost because it must be overhauled periodically. How about something with retractable landing gear? Sure it will fly faster, but there are things that can go wrong with a landing gear that moves up and down. How about the radios? More radios means more possibilities for something to go wrong.

If you have a growing family and more people to fly than seats to put them in, part of the decision has been made for you. If you use your airplane on business, you might find that more trips could be completed in less time if you had a 50-mph-faster airplane. (You might be able to coax a business partner to join you on some of the traveling to help share the expense.)

You've looked at all the practical aspects of this machine. Yes, you know it will cost more but you think the extra speed and payload capability will be worth it. Perhaps you've even decided to become partners with someone else you've met around the airport to share expenses. With this idea in mind you might find yourself getting a bigger airplane for only a few dollars more than you paid for the small aircraft you've been flying. You not only want this airplane, you've decided that you need this airplane and you believe the costs can be worked out.

As we worked out the operating costs per hour on the small single-engine airplanes in Chapter 4, you should pick up some copies of *Trade-A-Plane*. Consider what your budget will allow; these days, $50,000 could buy an early model Beechcraft Bonanza or a late model Cessna Cardinal very easily. If you want something even bigger, faster or newer in the single-engine line, be prepared to spend bigger dollars. A late model Cessna 210 could run $100,000. A student of mine recently picked up a 1986 A36 Bonanza and spent nearly $200,000 in the process.

If you've decided to buy something bigger and faster, you will find many airplanes: Beechcraft Sierra or Bonanza, Cessna Cardinal RG (retractable gear) and 210, Piper Arrow and Lance and Comanche, Mooney 201, Rockwell Commander, and Bellanca Viking.

SELLING AN AIRPLANE

Most pilots cannot maintain two airplanes at the same time. Before the new airplane can be swung into the hangar or on to the tie down, the old one will have to be sold. There are many different ways this can be done.

First, if you bought the airplane from a dealer, you could return and see if they'd be interested in buying back the aircraft. If the airplane is in good condition, this might be an easy option because a good used airplane is easier to sell. The dealer may not want to buy an aircraft outright but might find your next airplane and take your old airplane in trade. Realize that a dealer will be offering wholesale value on your airplane and a retail price on the next airplane.

The downside is losing money; the upside to the transaction is that all you have to do is taxi an old aircraft in and taxi the new one away after you sign the papers. It's this convenience that most people want and are willing to pay for.

You need to know the amount of money involved in a typical transaction to make a good decision on whether to offer the aircraft to a dealer or try to sell the airplane yourself. Let's assume the Cessna 172 you purchased has given all it can and you've decided to trade up to a turbocharged Cessna 210. For our paperwork process here, we'll assume the 172 was a 1971 model and that you paid approximately the going price for an IFR-equipped version, which was about $20,000. The next airplane you chose is a 1979 turbocharged Cessna 210, N model, which on the retail market is worth about $75,000.

When the dealer takes the 172 in trade, you will receive only the wholesale price for the airplane, about $15,000 toward the purchase of the next airplane. If the dealer sells you the aircraft at $75,000, that means he paid wholesale for the 210 also, which in this case is about $60,000. All wrapped up, it means you lost $5000 on your 172 and paid about $15,000 over wholesale to purchase the 210, giving the dealer approximately a $20,000 profit. That's quite a chunk of money to pay for paperwork.

Another method of selling an aircraft is do it yourself. You put a couple of FOR SALE signs giving your phone number in the windows of your plane at the airport. When someone calls, use your friendliest voice and try to have the person join you at the airport to see the aircraft and discuss the particulars. If you intend trying to sell the aircraft yourself, subscribe to *Trade-A-Plane* for an idea of what your aircraft is worth. This tabloid is another place to advertise the aircraft.

If the potential purchaser is a new pilot, look in the pages of *Trade-A-Plane* to find the names of banks that will finance your aircraft for the new owner. Local banks in the purchaser's area may also be good candidates, but the buyer will have to complete some of that legwork. The idea is to help the potential buyer over as many hurdles as possible. If the buyer agrees to purchase the airplane, have an attorney who understands aircraft draw up a bill of sale to explain any contingencies of the sale, such as mechanic's inspection and title search.

How much are you willing to pay if a mechanic finds problems with your airplane? To avoid hassles, this is something you should decide before you sign the agreement. Most of all, be truthful with the buyer. Tell how much you've enjoyed aircraft ownership, and why you want to sell this aircraft. Tell the new owner what to expect and what not to expect from the airplane. A little honesty now could save a great deal of work later.

Most banks require a title search to ensure that no liens are assigned against the title. Once the aircraft is determined to be in good mechanical shape and within its annual inspection period, the airplane can be sold with a transfer of registration and a transfer of funds. Selling an aircraft is much like selling a used car, so you must keep your plane looking good. No matter how well an airplane is equipped, most people believe the old saying that if the outside is poorly maintained the inside most likely is poorly maintained. That can squash a deal very quickly.

You also need patience to sell an airplane. Most airplanes don't sell the first weekend and it might even take months.

No matter how long it takes, you must be aggressive in your marketing techniques. Make sure those signs can be seen in your airplane. Run an ad in *Trade-A-Plane* and the local newspaper. Copy some flyers or index cards about the aircraft and fly around to all the airports within 50 or 75 miles and put copies on the bulletin boards. Tell people they can call you collect. Try sitting around the aircraft on the weekends so people can talk to you if they're interested. Hang plastic flags that real estate agents use on open houses to attract attention to your plane in the tiedown or in front of the hangar. The point is, the harder you work, the better your chances of selling the aircraft.

Once you've sold the aircraft, you can begin looking for the next one. Chances are by this time you've already had your eye on a particular model. If you're an AOPA member, you can call their library reference section and request copies of back issues of aviation magazines that carried stories on the plane you're looking for. You'll learn some interesting facts about the aircraft you are considering, both pro and con.

Walk around local airports and talk to people who own an airplane like the one you're considering. Ask them what they think about it. Is it expensive to maintain? How much does it really cost to fly per hour? Would they buy one again? This time spent could save a great deal of money in the long run. Certainly, don't let your emotions make a deal you don't feel comfortable about.

You're bound to find a few of the aircraft you're looking for in *Trade-A-Plane* or on the bulletin boards of some of the local airports. With the information you have already gleaned from other owners and the magazine articles you've read, you'll find it a bit easier to negotiate with a seller for the model you're seeking. Additional product information can be ordered over the phone from the AOPA library (if you're a member).

Maybe you will be able to pick up that Cessna 210 for $71,000 instead of $75,000. So your work selling the 172 and finding the 210 yourself could save $9000. That means that if your time is worth $50 per hour, there are 180 hours of work you would have to put into selling and buying of these two airplanes: the equivalent of nine working weeks, eight hours per day, and 40 hours per week before you are really losing any money. It's your decision and your money.

COMPLEX AIRPLANES

This is a list of some of my favorites and by no means includes all the retractable gear, constant-speed propeller airplanes available. These are not specific recommendations, but merely personal observations of particular aircraft through flying them over the years.

PIPER ARROW

If your primary training was in a Piper Cherokee or Warrior, the next step up in the line could easily be the Arrow (Fig. 8-1). If you stand next to one, the long clean lines of the fuselage combined with the standard fat Piper wing will make you feel like you've run across the older brother of an old friend. The advantages of the Arrow over its smaller siblings are numerous. With a 200-hp engine, variable pitch propeller, and retractable landing gear, you can carry four people, full fuel, and some baggage at speeds up to 150 knots. And all this with an airplane that burns only 12 gallons of fuel per hour from the two 25-gallon fuel tanks.

Glenn Chatfield

Fig. 8-1. Piper Arrow.

Why anyone would want to look beyond an Arrow after a Cherokee is anyone's guess. Once you've become comfortable using the retractable landing gear and the constant-speed prop, the rest of the checkout is a piece of cake. One reason is that Piper set up the cockpit in the Arrow to look almost like a carbon copy of the fixed-gear Cherokee. All the handles and switches and dials and circuit breakers in the Arrow can be found by looking in the same place as the Cherokee. Some models of the Arrow have an automatic gear extension system that virtually eliminates the chance of landing without putting the gear down. Newer Arrows come with a turbocharged powerplant to produce more power at altitude.

Piper Arrow performance data

Wingspan 32 ft
Length 24.6 ft
Height 8 ft
Engine Lycoming IO-360
TBO 1400 hrs
Empty weight 1504 lbs

Gross weight   2650 lbs
Useful load   1146 lbs
Fuel capacity   50 gal
Baggage capacity   200 lbs
Seats 4
Rate of climb   900 fpm
Service ceiling   15,000 ft
Maximum speed   175 mph (152 kts)
Normal cruise speed   165 mph (143 kts)
Range   683 statute miles (595 nautical miles)
1993 prices of used Piper Arrows $25,000 to $80,000

CESSNA CARDINAL RG (RETRACTABLE GEAR)

My first experience with a Cessna Cardinal RG was flying with an instrument student. The airplane had been around the airport for a year or so and I had grown accustomed to seeing it parked just a few rows down from my airplane. The Cardinal flew often. My checkout in the aircraft before the instrument training showed why it was a stranger to the tiedown.

The Cardinal was a quick airplane, about 140 knots at lower altitudes. It was also quick on the controls, unlike some of the Pipers flying with a big fat wing. Aileron control was slick enough that it made me wonder how many people had tried an aileron roll in the airplane, despite the illegalities.

The instrument panel on the Cardinal is wide enough for numerous radios and other equipment. The airplane I flew was equipped with ARC (Aircraft Radio Corp.) avionics that caused very little trouble during the training (except the transponder) although many pilots detest ARC quality. The Cardinal RG entered production in 1971 and during an eight-year run, nearly 1400 of them were built.

Cessna hoped the initial appeal of the machines would be the lack of visible wing struts, the retractable landing gear and the extra-wide doors. Those were indeed some of the items I found appealing, although there were problems with the early Cardinal RG landing gear. I remember many times after takeoff, the gear pump would keep running, trying to retract the gear after it had already done so. It would be necessary to pull the pump motor's circuit breaker to prevent burnout.

Other times we could not get a "down-and-locked" indication. In this particular model, Cessna used an amber light for the "in-transit" position and a single green light for the "down-and-locked" position. A quick look out the windows would indicate if the mains were down and the green light indicated OK on the nose gear. Later modifications on the RG fixed most of those gear problems. If you choose one, be sure it has the gear mods completed and that it has no damage history from a gear-up landing.

The Cardinal is another of those great machines that can be loaded up and flown long distances. The useful load is often more than 1000 pounds, which offers enough room to load four adults, full fuel, and some baggage to fly nearly four IFR hours. That's pretty hard to beat. On one long cross-country trip in the RG, I flew with two passengers and baggage from Chicago to Colorado Springs in just under six hours with only a single fuel stop. The airplane is roomy enough to make that trip comfortable to all except a tall passenger.

Even with the upsurge in used aircraft prices, the Cardinal has always sold for a lower price than the competition, the Piper Arrow and Beech Sierra. If purchased in good condition, a pilot can receive a great deal of airplane for the money.

Cardinal RG performance data

Wingspan   35 ft 6 in
Length   27 ft 3 in
Height   8 ft 7 in
Engine   IO-360 Lycoming (200 hp)
TBO   1400 to 1800 hrs
Empty weight   1765 lbs
Gross weight   2800 lbs
Useful load   1035 lbs
Fuel capacity   60 gal
Baggage capacity   120 lbs
Seats 4
Takeoff distance   890 ft
Landing distance   730 ft
Maximum speed   156 kts
Normal cruise   144 kts
Service ceiling   17,100 ft
1993 prices of used Cessna Cardinal RGs $24,000 to $46,000

BEECHCRAFT SIERRA

If you've never had the opportunity to fly one of the Beechcraft Sierra retractables, make it a point to look this airplane over before you buy. Beech has a history of building airplanes with the solid feel of a heavier machine, not to mention great interiors with very little cheap plastic. The Sierra (Fig. 8-2) uses the same 200-hp Lycoming engine that is in the Arrow and Cardinal. You can expect fuel flow to be about the same as those airplanes, about 11.5 to 12 gallons per hour. The Sierra also wins points from most pilots on the size of the cabin: wide and, unlike the Arrow, equipped with a door on either side to make entry easier for all. There's plenty of leg room in the front and back.

Once airborne, the noise level is the quietest of the Arrow-Cardinal-Sierra group. In cruise flight you can speak to the person next to you without yelling over the noise of the engine or the noise of wind whipping through any air leaks because the aircraft doors fit tightly. Certainly the roomy interior and the quiet ride come at a cost, part of which seems to be general performance; the 200-hp Lycoming Cardinal produces 144 knots and the Sierra will only do 135 knots. The Sierra is a dream to fly with controls that are incredibly responsive, but there have been many complaints about hard landings in the Sierra. In the landing flare, as well as on the takeoff roll, the Sierra requires the pilot to fly the airplane and make

Fig. 8-2. Beechcraft Sierra.

positive rotation unlike the Cardinal and Arrow, which practically fly off the runway. This can cause inexperienced pilots some grief. A thorough checkout before the pilot flies off on his own should be sufficient.

Beech Sierra performance data

Length   25 ft 9 in
Height   8 ft 1 in
Wingspan   32 ft 9 in
Engine   Lycoming IO-360 (200 hp)
Maximum gross weight   2758 lbs
Empty weight   1720 lbs
Useful load   1038 lbs
Fuel capacity   57.2 gal
Baggage capacity   270 lbs
Seats 4
Takeoff distance   1100 ft
Landing distance   850 ft
Maximum speed   168 kts
Normal cruise   135 kts
Service ceiling   15,385 ft
1993 prices on used Beech Sierras $33,000 to $55,000

MOONEY 201

By this point you are probably tired of hearing about aircraft propelled by the IO-360 Lycoming engine. There is only one more, and this one could be worth the wait. The Mooney 201 is derived from a basic design that is nearly 40 years old. What makes the Mooney such a spectacular airplane is that with the same engine used in the Beech Sierra, Cessna Cardinal, and Piper Arrow, this airplane will cruise at 165 knots. That's a full 30 knots faster than the Beech Sierra.

The natural question: "What did they leave out?" The answer is very simple: the drag. If you stand next to a Mooney, you'll understand very quickly what this means. The Mooney looks like a flashy Italian sports car. In fact, it looks like it could easily do 100 just sitting on the ramp. The engineers at Mooney decided to improve the performance of the Mooney the hard way, by earning the extra numbers through engineering improvements and not by mounting a bigger engine on the aircraft. Those Mooney people reduced drag everywhere possible; they improved the engine airflow with an efficient cowling; they reduced drag by completely enclosing the landing gear after it was retracted; they covered up the hinges for the flaps; they even boosted speed by eliminating the old fashioned outside air temperature (OAT) probe that stuck out in the slipstream like a rubber-tipped dart had been shot at the windshield. The results were certainly worth the work.

Inside, the Mooney 201 has a sports-car look that seems form fitting, but yet comfortable for all except the rear seat passenger behind a very tall pilot. Quick on the controls, the Mooney is pure heaven to fly, but requires attention from the pilot. The Mooney is not the kind of airplane you can turn downwind at cruise power and expect to put it on the numbers without plenty of room to slow down. The Mooney just likes to go fast.

A Mooney's load carrying capacity is not spectacular but very few pilots fly an airplane with all the seats full, so this should be only a minor inconvenience. It's overcome by the above-average speed and low fuel flow.

The popularity of this efficient airplane shows up in a higher than average price for a 200-hp aircraft. If you also were to ask why the other aircraft manufacturers don't clean up their aircraft aerodynamically the way Mooney has, I'd say that's a question that Beech, Piper, and Cessna have yet to answer.

Mooney 201 performance data

Length   24 ft 8 in
Height   8 ft 4 in
Wingspan   36 ft 1 in
Engine   IO-360 Lycoming (200 hp)
TBO   1800 hours
Seats 4
Gross weight   2740 lbs
Empty weight   1671 lbs
Useful load   1069 lbs
Fuel capacity   64 gal
Baggage capacity   120 lbs
Takeoff distance   850 ft
Landing distance   920 ft
Maximum speed   176 kts
Normal cruising speed   168 kts
Service ceiling   18,800 ft
1993 prices of used Mooney 201s $50,000 to $120,000

CESSNA 210

Cessna Aircraft Company began producing the high-performance single-engine Cessna 210 in 1959. Over the years it evolved from a four-place into a six-place aircraft capable of cruise speeds just under 200 mph. The 210 is powered by a 300-hp engine to achieve these results (Fig. 8-3).

Fig. 8-3. Cessna 210.

The Cessna 210 is the most popular of the six place, high-performance singles produced for general aviation. Very often, as with the Beechcraft Bonanza and the Piper Lance, a 210 was considered an alternative to buying a twin-engine aircraft. The 210, fully fueled, can carry six full-size adults and approximately 5.5 flying hours of fuel (much more than most passengers could stand anyway). If you were to lighten the fuel load for a bit less range, the aircraft could carry baggage for those six people. This performance is comparable to what you would expect from a twin-engine Cessna 310 or a Beechcraft Baron.

The 210 is heavy aircraft on the elevators and to a relatively new pilot could feel like a DC-3. The 210 is quite responsive on the ailerons, however, and is truly a great airplane to fly. If you fly your aircraft the way most people do, without every seat filled, the rate of climb can be spectacular, often approaching 1000 feet per minute. In IFR conditions this can push you up through an icing layer fast enough for it not to be a problem.

To aid in climb and cruise performance, many of the post-1966 models came equipped with turbochargers to maintain a constant manifold pressure. The major hangup about 210s was on the pre-1970 Cessna models that had many expensive landing gear problems. Post 1970 Cessna 210s have the larger six-place cabin. Earlier models carried only four people.

The 210 is a large, fast airplane that carries a good load. Late model 210s can be equipped with weather radar in an external wing pod.

Cessna 210 performance data

Length   28 ft 3 in
Height   9 ft 8 in
Wingspan   36 ft 9 in
Engine   IO-520 Continental rated 300 hp takeoff (285 hp cruise)
Gross weight   3800 lbs
Empty weight   2170 lbs
Useful load   1630 lbs
Fuel capacity   89 gal
Baggage capacity   240 lbs
Seats 4 to 6
Maximum speed   175 kts
Normal cruise   171 kts
Takeoff roll   1250 ft
Landing roll   765 ft
Rate of climb   860 fpm
Service ceiling   15,500 ft
1993 prices of used Cessna 210s normally aspirated $25,000 to $200,000
    Turbocharged $33,000 to $210,000

PIPER LANCE

The Piper Lance (Fig. 8-4) is sometimes called a station wagon with wings. You enter from the front or rear door and realize you're climbing into a vehicle with a vast amount of interior space. As competition for the Cessna 210 and Beech Bonanza, the Lance didn't fare too well, however. Powered with a 300-hp engine, you might have expected the Lance to be a real threat to the rivals. Unfortunately, to make the Lance run at speeds comparable to the 210 and the Bonanza, the Lance pilot must keep the throttle pretty far forward, thereby burning more fuel. On a normally-aspirated Lance, a pilot can expect to burn almost 22 gallons per hour to achieve the same 168 knots that a Cessna 210 pilot would use 16.5 gallons per hour to achieve. This can make a sizable difference in operating costs, although the purchase price of the Lance is much lower than the 210 or Bonanza.

The bargain that a Lance pilot receives is an airplane that is easy to load with people and cargo from ground level. The Lance also offers an optional seventh seat, which is just right for a small child because it is sitting on the armrest between two other rear seats. There is no doubt the Lance is heavy on the controls—all the controls. If the airplane is full of fuel with two large people in the front seats, the pressure on the stabilator during the flare to landing can be very heavy. On takeoff too, the ground roll of the Lance to clear an obstacle at the end of the runway can be long. Even longer if the airplane is loaded to the forward end of the center of gravity envelope.

In practical terms, the Lance will carry six people and a respectable three hours of fuel. The Lance has a lower purchase price initially. It also gains points be-

Fig. 8-4. Piper Lance.

cause the big Lycoming will run to a TBO of 1800 hours on the turbocharged version, much higher than either the Cessna or the Beech. Like its smaller brother, the Arrow, the Lance carries the advantage of an automatic gear extension system to help the pilot avoid an embarrassing arrival at the airport. The Lance also wins because the pilot doesn't need the seat cranked up to the top to be able to see over a high instrument panel, like in a 210 or a Bonanza.

Piper Lance performance data

Length   27 ft 8.2 in
Height   9 ft
Wingspan   32 ft 9.8 in
Engine   IO-540 Lycoming (300 hp)
Cruising speed   153 kts
Maximum gross weight   3600 lbs
Empty weight   1980 lbs
Useful load   1620 lbs
Baggage capacity   200 lbs
Seats 6 to 7
Fuel capacity   98 gal
Takeoff distance   960 ft
Landing distance   880 ft
Service ceiling   14,600 ft
1993 price of used Piper Lance (not turbo charged) $45,000 to $65,000

BEECHCRAFT BONANZA

If you asked which complex airplane was the most expensive among airplanes mentioned in this book, the answer would be Bonanza. Beechcraft Bonanzas have always been more expensive to purchase than their Cessna 210 competition. Like the Mercedes-Benz, however, the Bonanza (Fig. 8-5) has one of the highest resale values in the used single-engine market. There's a reason for this. Beechcraft Bonanzas stem from a long heritage that stretches back to the late 1940s when Bonanza production began.

Fig. 8-5. V-tail Bonanza.

Beech Aircraft Corp.

The Beechcraft Bonanza is one airplane that has formed a niche in the public eye, much like the Piper Cub and the Learjet. The Bonanza is known as "that airplane with the V-tail." Thousands of Bonanzas were produced with the distinctive V-tail over the last 35 years, but as of this date, Beech no longer produces that model. However, in place of the V-tail Bonanza, Beech has other Bonanza models with a conventional straight tail that range from the four place F-33 to the roomier six-place A-36. A turbocharged model is also available in two forms, the A- and the B-36TC Bonanza.

The F-33 has a single right side door for entry to the interior's four seats. The A-36 model has one door for the pilot and front seat passenger and a second double door in the rear to admit passengers into the club seating for four. The A-36 rear cabin also has a fold-down wooden table like the Bonanza's big brother, the twin-engine B-58 Baron.

Beech Aircraft Corporation's marketing strategy has always been to produce a smaller quantity of aircraft that reflect a "hands-on" manufacturing technique. There is a noticeable lack of plastic bits and pieces in a Beech Bonanza, as well as a more elegant interior, complete with either leathers or crushed velour.

That strategy can be a tremendous benefit if you buy a used Bonanza, once you pay the price. As the public knows about Cubs and Learjets, the people you fly with will know the Bonanza. Keep in mind that the Bonanza image is partially that, an image. When you check out the specs on the airplane later on, you'll notice it lacking in roominess and payload capability. That is just part of the deal. If you want a Bonanza, you must realize that it is not the greatest all-around airplane in the world.

Why would anyone pay more money for an airplane that isn't quite as fast as some nor as roomy as others? Like the Mercedes-Benz, the Bonanza is for the person that demands quality. You have only to look at a Bonanza to see the difference outside and in. The proof really comes when you fly the Bonanza. Nothing—and I mean nothing—flies like a Bonanza. It has the heaviness of a twin-engine airplane yet is responsive like a single. Even if you're not as smooth on the controls as you might be, the Bonanza will be. It doesn't fly through the air, it glides, almost effortlessly.

Before you buy any heavy single-engine airplane, fly a Bonanza. It is the benchmark by which other singles are measured.

Beechcraft Bonanza performance data

Length   26 ft 8 in
Height   8 ft 5 in
Wingspan   33 ft 6 in
Engine   IO-520 Continental (285 hp)
TBO   1500 hrs
Normal cruising speed   172 kts
Seats 4 to 6
Baggage capacity   270 lbs (F-33)
Fuel capacity   74 gal
Takeoff distance   1002 ft
Landing distance   763 ft
Maximum weight   3412 lbs
Empty weight   2279 lbs
Useful load   1133 lbs
Service ceiling   17,858 ft
1993 prices of used Beechcraft Bonanzas $20,000 to $300,000

MEMORY LOGBOOK

It's probably about this time that you are again beginning to believe that you know it all. I know that I've felt like that when it comes to flying. Let me share something that helped bring me back to earth.

PILOT'S DELIGHT[1]

I was to be flown from Palwaukee Airport near Chicago to a small grass strip near Rockford, Illinois, some 50 miles away, to pick up a friend's Cessna 150. The trip would be short, but by the time Steve, the 150's owner, and I were ready to leave, it was obvious that it would be near dusk upon our arrival at the unlighted field.

---

[1]"Pilot's Delight" reprinted from *AOPA Pilot*, October 1978, all rights reserved.

Considering the huge trees that surrounded the runway at our destination, Steve said he would rather drop me off at Rockford Airport where I would catch another ride, rather than risk landing at an unfamiliar field with darkness approaching. I smiled as Steve announced his lack of interest in completing the trip, but then he was a private pilot with only some 90 or so hours and I was an instructor with nearly 1500.

The sun was nearing the horizon when Steve let me out at Rockford and as my eyes searched the ramp they spotted one of the roughest looking Tri-Pacers I had ever seen. When I asked, the pilot of the old Piper told me that he indeed was to be my taxi for the five-minute flight to the grass strip.

I gave the bird a quick once-over, raised an eyebrow rather skeptically at its appearance and climbed in. "Let's get going," I said. I knew the approaching darkness wouldn't be a problem if we wasted no time getting airborne.

Airborne again, I could see just how beautiful a midwestern sunset could be with the streaks of pink against the now graying sky. It seemed, well, almost peaceful looking. I thought of the old rhyme for a moment: "Red sky in the morning, sailors take warning, red sky at night, sailor's delight." It almost could have been written for pilots, too.

My mind returned to matters of the moment as Jack, the Tri-Pacer's pilot, announced our arrival over the strip. It still looked rather light in the air, but as my eyes moved toward the ground I became aware of just how dark it really was becoming.

I mentioned uneasily that perhaps we should go back to Rockford and return to the strip by car, but with a casual wave of his hand, Jack assured me that he was experienced at this sort of thing. Sometimes the person with more experience must take control of a situation, I thought. It made sense, so tonight was my turn to be the student.

The plan had been to fly over the top of the strip until someone on the ground turned on some auto headlights to aid the pilot in his approach, but as I gazed out the window, blackness spotted by an occasional streetlight was all that met my eyes.

Jack looked over at me and smiled as he turned base for what I still believed to be an imaginary runway. Returning his smile rather weakly, I started thinking that we should call the whole thing off, but then Jack was more experienced in these matters than I.

We were now on final and, as we approached closer, I could see the lights of a car—one car! My eyes moved quickly to the altimeter and I saw that we were but 200 feet or so from the ground. Another quick glance out of the window and I was aware that the silhouettes I saw against the sky were the huge trees that surrounded the field. But the tops were now above us!

With sweat pouring from my brow and with just one set of lights on the ground, my depth perception was almost nil. If I didn't do something quickly, Jack and I were going to be tomorrow's headlines. Instinctively, my hand went for the throttle. "No!" Jack shouted as he yanked it back to idle. The airplane sank immediately. We hit the ground so hard that I was convinced my life had but a fraction of a second left in it.

When I realized a few seconds later that I wasn't dead, I looked out the window to see if the wheels were still attached. Almost immediately I returned my gaze to the inside of the airplane when I realized I couldn't see the landing gear for the darkness. We rolled to a stop at the end of the runway, but not even waiting for the engine to die, I jumped out and gave a terrific sigh of relief that somehow I had managed to survive this insane journey.

That little voice inside my head had tried to tell me the trip was foolish, but I'd refused to listen. Never, never again was I going to let another pilot, experienced or not, put me in a situation that my common sense tells me is dangerous.

As I spoke later with Jack about the state of my nerves, I asked just how much time he'd logged. "Oh, let me see now," he said as he scratched his head, "must be about 70 or 80 hours by now."

Interesting answer I thought, but the best was yet to come. "Jack," I said, "How many times have you landed at this strip at night?"

"This is my second," Jack replied with a big grin, "How'd I do?"

9

Building your
own airplane

I GREW UP IN CHICAGO and during my teenage years an interest in aviation could be satisfied two ways: watching real airplanes and building model airplanes.

As a kid, I built hundreds of model airplanes, some from plastic kits, but most from good ol' wood, metal, and the airplane dope used to paint them. In later years, some of the models actually flew, even a few I designed myself. In no time at all, I was conversant with the language of aircraft construction; ribs and wing spars as well as jigs and longerons. It was a wonderful experience, but I knew full well that all this work would have nothing to do with real flying airplanes. After all, this is how you built model airplanes.

Was I ever wrong!

In the summer of 1964 I picked up a part-time job at the Greater Rockford, Illinois, Airport, parking airplanes for an air show. I'd never seen a real air show before, so I thought it would be a lot of fun. And it was. I wasn't prepared for some of the airplanes I saw taxi by on the way to parking spaces in the grass. I saw machines that looked like they had been built from kits: bits and pieces of wood and metal and fabric put together. These airplanes really flew!

One of the men I was working for told me that most of the people flying these airplanes had built them from the ground up with their two hands. I was stunned. Everywhere I looked, I saw many incredible examples of homebuilding techniques. Aircraft with interiors of rich leather and instrument panels of cherry wood with each instrument location carefully selected and installed. Outside the paint was a kaleidoscope of color, like starburst schemes with bright stripes. There were lemon yellow airplanes, and some lime green, and bright fire engine red. It was my model airplane shop coming to life.

That air show is no longer held at Rockford. The airport proved unable to accommodate all the aircraft, people, and automobiles. These days, the Experimental Aircraft Association's week long air show at Oshkosh, Wisconsin, has become even more world famous than the years when I worked with them. "Oshkosh" is usually the last few days of July and the first few days of August each year (Fig. 9-1). If you've never had the opportunity to visit the show at Oshkosh, it's well worth the trip. These days, you never know what you might see during the week. It could be the Air Force Thunderbirds, a B-1 bomber, a 747 that stops in for a look-see or perhaps even the supersonic Concorde.

Fig. 9-1. Experimental Aircraft Association convention at Oshkosh, Wisconsin.

CAN YOU REALLY BUILD IT YOURSELF?

Is it possible that there are others like myself who have dreamed of seeing an airplane come to life from the work of their own two hands? Maybe you've thought about it, but didn't think you had the skills for building a plane. Don't let that slow you down. In a recent article in *Sport Aviation*, EAA member Budd Davisson wrote: "Evaluating yourself as a builder has absolutely nothing to do with what you have done or what you can do. Skill and experience aren't part of the equation, although they do help . . . There is not one skill involved in building an airplane that can't be learned by the average man (or woman) at the airport who has the right attitude. And that's the magic ingredient . . . attitude."

Now the secret is out. If you want to, you really could build a machine of your own, piece by piece, until an airplane was the result. How do you know you want to build an airplane? How do you know your motives are right to involve yourself in this kind of a project?

Building an airplane is labor, although to most pilots it is a labor of love. With some of the newer kits available today many of the major parts are already assembled. You might be able to work a normal job and, using your spare time in the evenings and some of your weekends, have the machine flying in perhaps two to three years. Yes, years.

Building an airplane is an incredibly labor intensive job if the job is going to be completed properly. After all, you want to be proud of the airplane. If you're working under the self-imposed pressure to build an airplane so you'll have something to fly, the results are not going to be very impressive. There are techniques in building, such as running the plumbing, or the electrical harness, or fabric installation, that require not only a great deal of your time, but a great deal of patience to be certain everything is put together properly.

Forming parts from metal or being certain that each wing rib is sanded "just so" before it becomes a part of the finished wing is important, because those are some of the details you might casually skip over in an attempt to fly the aircraft as quickly as possible. When considering the need to produce a quality project just for your own sense of satisfaction, there is the even larger need to produce a safe aircraft.

Remember, you'll be trusting your life and the lives of your family and friends to this machine. You should have no higher goal in building an airplane than to produce the highest quality machine possible. If you want an airplane to fly right now, go out and buy one that's already built.

SCRATCH-BUILT OR KIT?

If you've read this far you must still be interested in building. Now all you must do is decide what kind of airplane you'd like to construct from the dozens that are available. There are really just two types of projects for you to consider. The first is an aircraft built from a set of plans that includes the all-important materials list.

After you look the plans over, you must find the proper materials yourself, every piece of wood and metal and fiberglass, every nut, bolt, and screw. This means you could spend weeks trying to locate the materials, not to mention the time involved in cutting and fitting those initial bits and pieces together. To some, this is what building is all about, starting from scratch and watching the lumber and aluminum become an airplane. Realize as you begin a project from scratch, that you're taking on a job that will most likely consume as much as five to six years.

When I began considering homebuilts, about the time I was 16 or 17, building from scratch was about the only way you could construct an airplane. In the 1990s, though, we've been blessed with an incredible number of new composite building materials to produce wings, fuselage, and tail plane faster than some older methods. This is your second option.

Composite aircraft are strong and light, but require incredibly complex machinery to form and cure the parts, something the homebuilder would be unable to perform. This is why many of the new homebuilts these days are produced from kits, just like model aircraft building. The boxes are delivered and everything you need to make the parts into an airplane is included. All you need are the tools and the time.

The popularity of kits today has helped to produce a greater number of finished aircraft each year. That's because kits help the builder reduce the total number of hours necessary to produce the aircraft (Fig. 9-2). The total reduction in building time has brought hundreds of new people into general aviation in their own airplanes.

Fig. 9-2. The Rotor-Way Exec helicopter is assembled from a kit.

Which will it be? A scratch-built airplane or a kit? The decision is yours. You need to sit down and analyze your life-style to decide how much time you can devote to this project.

WHICH AIRPLANE SHOULD YOU BUILD?

The answer to this question is very simple. Which aircraft do you like? Which one will fit into your time schedule? Which one will perform the way you want after construction? These questions are a necessary part of the process when building an airplane.

If you want an airplane in the next three years, then you are wasting your time starting with a scratch-built machine. It will just take too much time to build. How do you plan to use the airplane? If it's just going to be flown once or twice a week for an hour's worth of cloud patrol, you might find a low powered airplane like a Volksplane (with a Volkswagen engine), a Pober Pixey, or a Firestar could work just fine. If you intend to fly aerobatics in the machine, you'll need something like an Acro-Sport because only certain machines are designed and stressed for aerobatics. Amphibian kits are available if a land and water airplane is more your style.

What if you want a long-range cruise machine to carry three or four people at a high speed? The new composite materials have allowed aircraft designers to incorporate ideas into some of the kits that cannot be found in factory-built aircraft. For instance, the new Lancair 320 uses the same basic 160-hp engine from a Cessna 172. While that 160-hp pulls the 172 through the air at about 125 mph, the Lancair will top out at about 250. It also has a range of 1100 miles, which is approximately double that of the 172. The price is a bit more than what you'd pay for a used 172 these days, but not much.

If I had an unlimited budget and could choose any machine, my personal vote would go to the Lancair IV. This beauty (Fig. 9-3) is a four place homebuilt rocket capable of IFR and aerobatic flight with a top speed in the area of 300 knots when the turbocharged version cruises around 24,000 feet. A pressurized version is available too. With IFR reserves, this aircraft will remain aloft long enough to fly more than 1000 miles. The Lancair IV prototype set a National Aeronautic Association world's record on February 20, 1991, during a flight from San Francisco to Denver. Over the 958 mile course, the Lancair IV averaged a speed of 362.48 miles per hour. Yes!

Lancair International Inc.

Fig. 9-3. Homebuilts have certainly changed from the wood and fabric days.

The Lancair IV is built of predominantly carbon fiber in epoxy resin matrix, a composite material 2½ times stiffer than fiberglass composites. While this aircraft not only incorporates high aspect ratio ailerons to make the stick forces firm but comfortable at high speeds, the Lancair IV incorporates retractable landing gear and slotted Fowler flaps for good landing performance. Even better, if you needed to take your aircraft off airport, you could store it in your garage by simply removing the wings and towing the aircraft on a small trailer behind a pickup truck.

One of the truly amazing parts of buying a Lancair IV kit is that many of the parts are already formed. The main wing spars are already installed in the wings, for instance. The flaps are already machined and the landing gear truss box is already assembled. Let's get to the part now that many readers are wondering about. How much? Current 1993 prices list the basic Lancair IV airframe at $45,900 and $65,800 for a pressurized model. Engine, propeller, and radio gear are extra. If this seems expensive, go back and check the prices of used aircraft in the last chapter. And those aircraft won't even come close to the performance of a Lancair IV.

Other aircraft of this same class include the Falco and the Glasair. Questair makes the Venture, which Questair claims flies at 240 knots and carries two people. All of these aircraft come in kit form. Usually, the fuselage, tail surfaces, and wings are preformed. That still leaves the plumbing, electrical, and engine gear to be connected, as well as the basic pieces of fuselage to be assembled. This will take less time compared to other homebuilts.

Join the Experimental Aircraft Association (EAA) and locate the closest chapter. After you attend some of the meetings you'll start to get a pretty good handle on just what aircraft are being built in the area. You can talk to the people who are already constructing various aircraft. If someone isn't building the kind of airplane you have an interest in, someone can probably refer you to another chapter. There you might find someone with direct experience on the airplane you want.

You have a pretty good idea of what type of airplane you're going to build and you've arranged for the money to finance the project, savings, checkbook loan, or whatever. You've even reconciled yourself and your family to how much time will be left over for them once you begin the project. Have you thought about where you're going to build the airplane?

Some projects have been completed in a basement, but those quarters can be pretty cramped. A one-car garage is also going to be tight, but at least it will allow you to move easily in and out. When it comes to constructing an airplane, there's no such thing as too much space. I can't imagine beginning work on an airplane in anything less than a two-car garage, although I know of people who have.

While it is not necessary to own a complete shop of tools before you begin, realize you'll have to buy hammers, saws, drills, and workbenches before you begin (Fig. 9-4). "The importance of the right workshop and the right environment cannot be overestimated," according to Bud Davisson in *Sport Aviation*. "Two to four years of your life are going to be spent in this environment and its effect on you can actually make or break the project. If it is a dingy, leftover space between bicycles and garden rakes, where you are constantly fighting the elements, the family, and the spiders, it is an uphill battle to keep your spirits up. If it is a nicely organized, bright, well-lit area that is dedicated to you and your airplane, then the only obstacles are inside your head."

There aren't an awful lot of decisions left to make, except to buy the plans or the kit and get started. Each day that you work, you might walk away trying to figure out when it becomes an airplane. If you're welding a tube fuselage, you'll see something that looks like an airplane in just a few weeks. If you're working with

Fig. 9-4. Building an airplane from plans.

EAA/Jim Koepnick

metal, and you must form most all the parts individually, it will take a great deal longer. Every few days though, I would shoot some pictures of the project. Hang them up on the wall in your workroom. As the number of pictures increases, the progress of your airplane will become clearer. As time passes, you'll find yourself installing wings, wheels, engine, and instruments. Before you know it, your dream of an airplane, put together by you, will become a reality, if you don't quit. Remember, you fail only when you stop trying.

During various phases of the construction, your airplane will need to be inspected by someone from the FAA. This is to be certain the aircraft will meet the standards necessary for it to be safe to fly. If your airplane is fabric-covered, the FAA will inspect prior to the airplane being covered to make certain all the welding is up to standard. If the aircraft is a more complex metal or composite construction more or possibly less inspections may be necessary, depending upon the aircraft type.

The day will come, though, when all the construction will be complete. You'll have made the engine test runups to be certain there are no leaks anywhere. You'll also have checked all the electrical parts to make certain the systems are operating properly. If they are, you need to find someone with a truck to help you move the airplane from its home in your garage to the airport where it can do what it was designed and built for—flying.

The initial tests at the airport will include more ground engine runs for the FAA as well as taxi tests to be certain the brakes and controls are sound. Soon though, there will be nothing left to test. Every last screw, nut, bolt, and wire will have been checked and rechecked. It's time to fly. Depending upon the airplane as well as the experience of the pilot, you might want to test fly it. You might want someone else with more experience in that aircraft to try it first. The first test is usually just a high speed taxi test. Each time though, the test speed rises as the flight controls are checked with air across them until the airplane actually lifts into the air. It's flying! And you built it (Fig. 9-5)!

Fig. 9-5. Homebuilts come in many different sizes like this Super Parasol.

If you're the kind of person who never seems to get enough praise for their work, be sure and take your airplane to Oshkosh or the EAA's Sun & Fun Fly-Ins. Park it with the other airplanes of your type. You'll run into thousands of people more than willing to eye your machine and tell you what a great job they think you've done on the construction. You'll hear stories about how those pilots would like to build something like your plane someday, but just don't know where they'd find the time or the space. You'll be able to remember when you were in their shoes.

Building an airplane is something special. Maybe more special than flying. You've watched the airplane emerge from a pile of parts and now it has meaning and substance. It's not just any airplane, it's yours. The one you gave birth to. Never forget that. Now go out and have fun (Fig. 9-6).

Fig. 9-6. A Vari-EZ is pushed toward the flight line at Oshkosh.

10

Some final thoughts

THROUGHOUT THIS BOOK I've shared some of the fun, some of the romance, and some of the good feelings I get from flying (Fig. 10-1). If you've read this far in the book I must have succeeded. Flying is one of the few things you can do that actually allows your family or friends to become directly involved in your hobby or your job.

While I enjoy tennis, my wife seldom goes along to watch me chase that little ball around the court, nor do I find much pleasure in watching her work out in the aerobics class. Flying, though, is a hobby you can enjoy just for the fun of going aloft to bore holes in the sky once a week if you like.

EAA/Jim Koepnick

Fig. 10-1. Weekend visitors to the EAA Air Adventure Museum receive a special treat as they watch the antique and classic aircraft take off from Pioneer Airport.

Or, you can use your skills to transport you and your family to some exotic location for a weekend's fun, or to close a business deal and still make it home for dinner. If you really let your mind wander freely, the possibilities of how to use your pilot's license are almost endless.

I think I'd be lacking in my responsibility as a pilot, teacher and as a writer if I did not bring one area of your pilot license to light for you. If you pull your pilot license out of your wallet, you'll find that line IX says, "has been found to be properly qualified to exercise the privileges of Private Pilot (or commercial or airline transport pilot)." The word privilege is the really important part of your license.

You, as a pilot-in-command of any aircraft, must realize that the issuing authority for your license, the FAA, has the right to take away your flying privileges if you violate the FARs. This removal of your flying privileges could stem from an innocent misinterpretation of an air-traffic controller's instruction or a willful disregard for some other regulatory section of the rules.

Either way, you could be kept from flying by a mild 30-day suspension to a stronger 90- or 120-day suspension. You could, however, receive a complete revocation of your certificate, which could mean your flying privileges would be halted for good.

No one wants to see this happen to a pilot, especially another pilot, but flying has changed a great deal since I first learned in the mid-1960s. Then, air traffic was lighter around major airports. Now, with the deregulation of the airline industry in 1978, total air traffic movements have reached new highs. While this idea of increased air traffic at metropolitan areas might not seem too important if you live outside an urban area, it should be.

You as an airplane pilot are required to know the rules that govern the pilot and operation of the aircraft. Because enforcement actions increased in the late 1980s, more and more pilots have been finding out that the FAA is not kidding about enforcement of the rules. More and more, the FAA has been playing airborne traffic cop with the pilot as the victim they'll try to catch in their traps. While a suspension or revocation notice is always open to appeal to the National Transportation Safety Board, the NTSB Bar Association reported in mid-1993 that of the 179 cases on appeal in 1992, the pilot prevailed in only 14. In previous years, the numbers of winning pilots was considerably higher.

Don't let it happen. Realize that your ability to fly an airplane is a privilege and not a right. It's a privilege that must be protected and you're the person who must protect it. As you fly around the country you'll most likely find small and large towns to whom small aircraft seem more a nuisance than a blessing. These residents usually have no more of an idea the difference between a Piper Cherokee and a Cessna 172 than they do a Learjet and a Boeing 747.

Some small airports have even started night curfews against all air traffic, no matter how small. In case you're wondering why they did this, it's often because some goofy group of pilots made takeoffs over surrounding houses when it wasn't necessary. Perhaps they were seen doing an aerial buzz job on homes near the airport. The point is that these restrictions didn't appear by themselves. Some other pilot just didn't care about the people who live near the airport. Please, if you gain nothing else from this book, realize that you really are a part of a larger group of pilots. What you do or don't do in your airplane is going to reflect on the pilot population in general.

A NEW GAME PLAN

During the years after airline deregulation, stories began appearing in newspapers and magazines about a growing shortage of pilots qualified to fly the nation's air carrier fleets. There have been strong debates over the years about just how severe this shortage is. The most important thing to consider here is that the pilot who begins flying for the airlines came from somewhere, most likely from the cockpit of another airplane. If a pilot moved up to a Boeing 727, he might have come from a regional airline, which created an opening for his job. That will probably be filled by a pilot moving up from the ranks of a corporate pilot job. Each job that opens at the top opens another job for someone down below. Whether we have enough pilots to maintain our air transportation system into the next century will be determined by people like you.

You as a pilot can do more to foster the career plans of some future pilot than you might know or even believe (Fig. 10-2). The General Aviation Task Force was a group of aviation organizations formed in the late 1980s to inform and promote general aviation as a viable means of transportation, as well as a superior source of revenue to both large and small communities. While the original task force has been disbanded, a new program was begun in mid-1993 to take its place, called the Learn To Fly Program. Before you can help promote general aviation to a prospective pilot or your community in general, you need to know the facts.

Fig. 10-2. It's never too early to start a future pilot's education.

- The general aviation fleet totals 212,000 aircraft, representing more than 87 percent of the entire civil fleet. By contrast, major airlines operate approximately 4000 aircraft, while the military accounts for 27,000.

- General aviation aircraft are used for business transactions, medical emergencies, surveying, fire-fighting, recreational flying, law enforcement, transportation of government officials, and mail distribution.

- Among 12,000 landing facilities, general aviation serves about 97 percent of them. For every major airliner that takes off this year, 50 general aviation aircraft will fly. For every hour an airliner flies, general aviation aircraft will fly almost three.

- More than 250,000 jobs are directly connected to general aviation and thousands of others associated with the industry are provided nationwide with an annual payroll of more than $12.7 billion. These jobs include mechanics, line service technicians, pilots, and manufacturing and support staff.

- General aviation is a $15 billion a year industry, contributing more than $400 million in tax revenues to communities annually.

- General aviation air ambulances have transported more than 600,000 seriously ill or critically injured persons to hospitals and trauma facilities since 1972. In addition, more than 2500 heart and liver transplants were made possible in 1987 through transportation provided by general aviation aircraft. Today, 185 emergency response programs exist across the nation, utilizing nearly 240 general aviation rescue aircraft.

(Statistics provided by Federal Aviation Administration, Aircraft Owners and Pilots Association, National Business Aircraft Association, Future Aviation Professionals of America, General Aviation Manufacturers Association, National Air Transport Association and Helicopter Association International.)

The point of the original General Aviation Market Expansion Plan (GAME Plan) was to increase awareness, understanding and usage of general aviation services around the country. The GAME Plan also showed travelers how general aviation services can provide an efficient transportation alternative. The new Learn To Fly Program will continue to work with very much the same goals. To learn more, dial 1-800-I CAN FLY.

If you're a flying businessperson, no one need tell you about the advantages of being able to fly yourself where you need to go, when you need to go, without all the hassles involved in airline travel. Corporate fliers could support general aviation by telling business friends. Explain the facts and figures on airports served by the airlines and those served by general aviation. Currently, 70 percent of all airline passengers transit only the top 25 airline airports in the nation. Tell friends they could be saving time and money by looking into chartering an aircraft to carry those four executives on that 165 mile flight instead of driving or flying airline connections (Fig. 10-3). Tell friends that more than two-thirds of all business aircraft trips make use of reliever airports that are located much closer to business destinations than commercial air terminals.

Fig. 10-3. The corporate airplane is no longer for CEOs only.

Beech Aircraft Co.

Fig. 10-4. A Beech King Air 200 is frequently used for charter flights.

(Figures courtesy of National Business Aircraft Association, General Aviation Manufacturers Association, U.S. Travel Data Center-1987, and the General Aviation Task Force.)

Tell friends the names of the local air charter companies at your airport, as well as the name of the person to contact (Fig. 10-4).

A qualified person can explain how advantageous chartering an aircraft can be. You might be able to explain just how helpful owning an aircraft is for a small company or individual businessman. Try taking a few business friends on a trip around their territory one day to introduce them to general aviation and its benefits.

You as a pilot have a freedom enjoyed by few individuals, not just in the United States, but in the world. Flying is a privilege that must be protected. Join some of the aviation associations that interest you and make sure you support them. You're entering a wonderful world when you fly. Fly safely, but fly.

Hopefully for you this is not the end, but rather just the beginning.

Index

buying an airplane, 63-76
  advertisements, decoding ab-
    breviations in ads, 63-64
  altitude encoders, 65
  appreciating value of older air-
    craft, 77-78
  automatic direction finder
    (ADF), 65
  autopilots, 64
  complex airplanes, 134-144
  engine, 64
  exhaust gas temperature gauge,
    65
  fabric-covered airplanes, 78-79
  flying clubs, 75-76
  leaseback, 70-72
  marker beacons, 65
  operating costs, 66-70
  partnerships, 72-75
  points to look out for, 65-66
  pre-purchase inspection, 66
  radios, 64
  taildragger airplanes, 78-79
  think before you buy, 131-132
  title search, 133-134
  Trade-A-Plane, 65, 132, 133, 134
  transponders, 65
  used airplanes, 78-79

C
California Coast, 12
categories of aircraft, 24, 27-29
certified flight instructor (CFI),
  23, 24, 38-41
certified flight instructor (CFI) li-
  cense, 102-105, **104**
Cessna aircraft company, 77
Cessna 120, 90-91
Cessna 140, 90-91, **91**
Cessna 150, 32, 60, 79-80
Cessna 152, 32, 62, 77-80, **79**
Cessna 172, 60, **61**, 81-82, **81**
  cockpit, **10**
Cessna 182, 62
Cessna 210, 62, 140-141, **140**
Cessna 310, 32, **101**
Cessna 402, **25**
Cessna Cardinal RG, 136-137
Cessna Owner Organization, 123
charter pilots, 106
charts and maps, 10-11, 25
  cross-country flight, **96**
  visual flight rules (VFR) chart, **12**
Cherokee Pilots Association, 124
CIGAR checklist, 59
circuit breakers, 8
Citabria 7ECA, 78
Civil Air Patrol (CAP), 18, 115-
  116, **116**
Class B airspace, 95-96
Class C airspace, 95-96

Classen, Chuck, 19
classes of aircraft, 24, 27-29
clearance, 11
Commander 114B, **73**
  cockpit, **8**
commercial flying, making
  money at your hobby, 6
commercial pilot license, 24, 99-
  100
communications, 25, 49-52
  approach control contact, 52
  ground control contact, 51
  nontower airports, landing pro-
    cedures, 52
  phraseology of ATC, 51-52
  tower contact, 51
  weather information, 52
communications radio, 8-9
compass, 8
CompuServe, 125-126, **126**
computers, personal computers,
  125-128
Confederate Air Force (CAF),
  117-118, **118**
control tower, 35-36
control wheel or yoke, 9
copilot jobs, 106
Corporate Angel Network
  (CAN), 109
cost of flying, 3, 13, 36-38, 66-70
Crites, Dale, 6
crop dusting, 106
Cub Club, 124
Curtiss P-40 Warhawk, **118**
Curtiss Pusher "Silver Streak"
  (circa 1912), **6**

D
David E. Neumeister Aircraft
  Newsletters, 125
Davisson, Budd, 148, 152
delays, 5
demonstration rides, 3-4
Denver airport, 5
Disney World, 12
DUAT, 10

E
Earhart, Amelia, 4, 109
electrical switches of typical airplane, 8
elevators, 7, 9, 11
emergencies and emergency pre-
  paredness, 44-45
en route procedures, 11
engine controls of typical airplane, 9
engine failure, 44
engine gauges of typical airplane, 9
engine run-up, 49

engines, 26, 64
Englehart, E. Allan, 26, **27**
Enstrom helicopters, 32
exhaust gas temperature gauge,
  65
Experimental Aircraft Association (EAA), 18, 112-114, **113**,
  152, 154

F
fabric-covered airplanes, 78-79
Falco homebuilt, 152
fear of flying, 3
Federal Aviation Administration
  (FAA), 2, 29, 158
Federal Aviation Regulations
  (FARs), 26, 29, 158
fixed base operator (FBO), 33-35,
  34, 36-38
flight computers, 26
Flight Computing Catalog, 128
Flight Deck Software, 127
flight instructors, 106
flight instruments of typical airplane, 7-8
Flight Line, 119
flight office, airport, 33
flight plans, 10
flight reviews, 96-97
flight service station (FSS), 10
Flight Simulator computer program, 127
Flight Standards, 26
flight test, 53-55
Flying magazine, reprint of
  "Plateau of Arrogance," 58-60
flying-club ownership arrangements, 75-76
Frasca basic flight simulator, **4**
freelance instructors, 36-38
frequency of airplane radio
  transmission, 9
friends and flying, 16-21
fuel calculations, 10-11
fuel gauges, 9
fuel mixture control, 9
fuselage, 7

G
general aviation aircraft, 32
general aviation fleet of U.S., 160
General Aviation Market Expansion Plan (GAME Plan), 160
general aviation statistics, 160
General Aviation Task Force, 159
Glasair homebuilt, 152
gliders, 14, **16**, 24, 28, 30, 31, **31**
Greth, Phil, 19
ground schools, 38
Grumman F4 Wildcat, **118**
Grumman Trainer, 87-88, **87**

oil temperature gauge, 9
operating costs, 66-70
Orlando, Florida, 12
Oshkosh Fly-In of EAA, 112, 114, 147-148, **148**, 154

P

Palwaukee Airport lunch bunch, **17**
partnership ownership arrangements, 72-75
personal computers (*see also* simulators), 125-128
"Pilot's Delight," from *AOPA Pilot* magazine, 144-146
pilot-in-command (PIC), 97
Pilots International Association (PIA), 119-120
Piper aircraft company, 77
Piper Archer, 83
Piper Arrow, 135-136, **135**
Piper Cherokee, **64**, 82-83, **82**
Piper Cub, 88-89, **88**
Piper Lance, 141-142, **142**
Piper Owner Society, 123
Piper Seneca, **103**
Piper Supercub, 88-89
Piper Tomahawk, 32, 83-84, **84**
Piper Warrior, 32
Plane & Pilot, 119
"Plateau of Arrogance" from *Flying* magazine, 58-60
Poberezny, Paul, 114
pre-purchase inspection, 66
preflight inspections, 11, 45, 48-49
  CIGAR checklist, 59
private pilot license, 23-24, 29, 31-32, **56**
Private Pilot Practical Test Standards (PTS), 54
prohibited airspace, 95
propeller, 7, 9, 11
protected airspace, 95

Q

Questair Venture homebuilt, 152

R

radio navigation, 26
radios, 8-9, 64
ramp areas, 33

rate-of-turn indicator, 8
recreational pilot license, 24, 27
redundancy in airplane systems, 9-10
rental airplanes, 60-63
Replica Fighters Association, 124
Robertson, Cliff, 114
Robinson helicopters, 32
Rotor-Way Exec helicopter, **150**
rotorcraft, 24
rudder, 7, 9
run-up of engine, 49
runways, 33

S

safety of flight, 10
sailplanes (*see* gliders; soaring)
scheduling instruction time, 45-47
Schweizer 1-35 sailplane, **16**
Seaplane Landing Directory, 121
Seaplane Pilots Association (SPA), 121
seaplane, **121**
search and rescue, 10
Sedona, Arizona, airport, **36**
selling an airplane, 132-134
simulator, flight simulator, **4**
  Advanced Flight Trainer computer program, 127
  computer-program simulators, 126-127
  Flight Simulator by Microsoft, 127
  Instrument Flight Trainer computer program, 127
Skyways, 123
soaring, 14, 28
  Schweizer 1-35 sailplane, **16**
Soaring Society of America (SSA), 115
spark plugs, 8
Sport Aviation, 148, 152
stabilizers, 7
stalls, 44
starter system, 8
statistics on general aviation, 160
steering system, 9
Stits, Ray, 114
student pilot certificate, 23
Super Parasol homebuilt, **154**
suspension of license, 158

T

tachometer, 9
taildragger airplanes, 78-79
takeoff power, 9
takeoff procedures, 11
taxiways, 33
Taylorcraft, 89-90
Taylorcraft Owners Club, 124
Ten Greatest Lies...About Flying, 14-15
terminal control areas (TCA), 95-96
throttle, 9, 11
tie-down areas, 35
time calculations, 26
title search, 133-134
Trade-A-Plane, 65, 132, 133, 134
transponders, 9, 65

U

ultralights, 120
United States Pilots Association (USPA), 119
United States Ultralight Association, 120
used airplanes, buying guide, 78-79

V

vacuum gauge, 9
Vari-EZ homebuilt, **155**
vertical speed indicator, 8
video magazines, 129-130
visual flight rules (VFR), 24
  chart, **12**

W

Warbirds Division of EAA, 18, 114
Water Flying, 121
weather, 10, 11, 26, 45, 52
weight and balance, 26
Whirly-Girls, Inc., 122
wind sock, 35, **35**
wings, 7
World War I Aeroplanes organization, 122-123,
Wright Brothers, 114
WWI Aero, 123

Y

Young Eagles program, EAA, 114